AN ADVENTUROUS LIFE
A PERSONAL AND CULTURAL HISTORY

ROBERT HAUPTMAN

ANAPHORA LITERARY PRESS

TUCSON, ARIZONA

Anaphora Literary Press
5755 E. River Rd., #2201
Tucson, AZ 85750
www.anaphoraliterary.com

Book design by Anna Faktorovich, Ph.D.

Copyright © 2013 by Robert Hauptman

All rights reserved. No part of this book may be reproduced in any form or by any electronic or mechanical means, including information storage and retrieval systems, without permission in writing from Robert Hauptman. Writers are welcome to quote brief passages in their critical studies, as American copyright law dictates.

Cover Image: Bob Just Below Summit of Mt Adams, WA, in whiteout, by Frederic Hartemann.

Published in 2013 by Anaphora Literary Press

An Adventurous Life
Robert Hauptman—1st edition.

ISBN-13: 978-1-937536-40-4
ISBN-10: 1-937536-40-8

Library of Congress Control Number: 2013904196

AN ADVENTUROUS LIFE

———————

ROBERT HAUPTMAN

For Terry and Kira

CONTENTS

Preface: Ethics, Memoir, and an Orienting Philosophy	9
One: Beginnings	11
Two: Vermont	20
Three: Education	35
Four: Interests	44
Five: Reading and Music	53
Six: Sensitivity	61
Seven: Health	64
Eight: Work	68
Nine: Intellection	76
Ten: Travel	80
Eleven: Skiing and Mountaineering	94
Twelve: Representative Climbs	106
Thirteen: Friends	116
Fourteen: Lovers, Wives, and Daughters	122
Fifteen: Concluding Remarks	133
References	137
About the Author	139
Index	141

PREFACE
ETHICS, MEMOIR, AND AN ORIENTING PHILOSOPHY

An Adventurous Life is a true account of my experiences. It is neither falsified nor embellished. I am no Wilfred Thesiger, but my experiences have been excitingly expansive enough to preclude the necessity of exaggerating or distorting for effect. An account that claims to be autobiographical in nature should reflect reality. The theoretician who rhetorically asks why one's memoir should be constricted by the facts does herself, her publisher, and especially her readers a disservice. No one likes to be fooled. The overly imaginative should stick to fiction.

When I am unsure about a specific occurrence, location, or date, I make this clear. Even when dealing with my earliest childhood, I am fairly certain about my age; I am at least in the correct ballpark: I know the difference between Yankee Stadium and the Polo Grounds. I know that I was two, six, seven, or ten. I am objective. I have an excellent (unfettered) memory and recall a great deal. (For example, we left Knickerbocker Village (KV) when I was 12; I remember a vivid life there; my brother was six; he recalls virtually nothing.) And this increases in quantity and precision as I mature. By the time I reach ten, my memories are solid, replete, and bountiful. And because my experiences have been so unusual, I may recall more of, say, my thirteenth year, than some people know of their entire childhoods. In many cases, their years were symmetrical and repetitive. Some of mine were as different as if I had reached puberty living with a Sherpa family in the Solo Khumbu and then moved to San Francisco.

Throughout the course of my adult life, that is, from the age of 17 onward, I have cared primarily about women, classical music, reading, and nature. I do not list them in rank order because it would be impossible to give precedence to one or the other. I care about all four simultaneously, and my life would be much diminished were I forced to disengage from any one of them. I have spent a very high percentage of my time in pursuit of these interests and they turn up with some regularity here, even if unmentioned. It stands to reason that I also care a great deal about animal rights, human freedom, justice, truth, peace, and many social issues, but I take these for granted and do not harp upon them. Indeed, they are all subsumed under the reading rubric as are my many generally unrecorded activities, e.g., my obligation to indicate to a senator representing each of the 50 states that the Viet Nam conflagration was a disaster and that we should leave immediately. We eventually did, though we continue to make the same type of blunder.

I have anonymized a few names: When I use an initial, I mean to protect a person by disguising his or her identity. All other names reflect reality.

ONE
BEGINNINGS

I teeter precariously on the lip of a 70 degree couloir. I have finished adjusting my crampons and I foolishly stamp my feet to set them in the snow. Fred is nearby and we are ready to start down along a half mile of steep, icy terrain. Suddenly Fred is gone: He is tumbling. There are no impediments and so he continues unabated. I watch in horror. He does not appear to be attempting to self-arrest, but I am in such a mesmerized state and he is moving so quickly that I may misperceive what is occurring. After many hundreds of yards, he manages to get on his stomach, with his head uphill, and thus in a position to begin to drag his ice ax along or plunge it into the icy snow. He stops. Had he continued all the way to the bottom, where this South Teton couloir meets the moraine of Garnet Canyon, he would have crashed into rocks, and that would have been a tragedy. He is far away and I have no idea whether he is okay, but I begin to descend, side-stepping extremely slowly. It takes a long time for me to reach him and I am physically and emotionally drained. We continue to the bottom, where I beg to rest for a few moments, but Fred refuses because the thunder and lightning are increasing and he wants to avoid a strike such as the one we learned of the night before: As we had relaxed in our motel room, the television announcer informed us that two people had been hit by lightning. We crawl downward, and Fred regales me with the fact that in his hundreds of Alpine ascents in France, Italy, Switzerland, the US, Mexico, and Canada, he had never fallen. He also informs me that he will never climb again.

A year later, we are back: This time we stroll up the Middle Teton along with many inappropriately attired tourists on a lovely, sunny day. We look across at our couloir and are both stunned to see that it is much steeper than we had thought. We were very lucky. After many years of climbing, one might have assumed, with some justification that we would have roped up before ascending, but it never occurred to us. On the next technical climb we did, a traverse of Cirque, we eventually roped up and proceeded with the kind of scrupulous caution one manifests on Everest's Hillary Step.

I grew up on the Lower East Side of New York City, just below Chinatown, in Knickerbocker Village (KV), the first major apartment complex built in the United States. It covers a large city block, comprises two sets of 6 interconnected buildings with some 1600 apartments. The entire enormous complex is undergirded by a labyrinthine basement with many twistings and turnings. In memory, the basement remains a magical place, where I walked, ran, or careened along on a bicycle. Only occasionally did one meet another person and then caution was warranted. KV is just a few blocks from what was then the Fulton Street Fish Market, now, in a sense, replaced by the South Street Seaport. The latter is a tourist attraction; the former, like the meat market to the north, was a working enclave of wholesale fishmongers. It opened very early in the morning and produced a volatile stench that wafted its way along South Street and north as well, probably infusing Wall Street with notice of its existence.

Even in these early days, I must have liked climbing because one day when I was six or seven, some busybody looked out of her window, saw me working my way along the lips that the jutting bricks offered high above a deep concrete pit, and telephoned my mother, who must have fainted. I do not know whether she raced out; the complex was so large that it could have taken her a half hour to locate me. Had I fallen, I would have been seriously hurt or killed. I did not learn my lesson, for many years later, I was on my way to my parents' summer home in Southern Vermont. Near Bear Mountain, my friends went into a diner for breakfast, but I decided to climb a cliff. This was a mistake because I got stuck and it took some fearful work to extricate myself from this unpleasant situation. If I mentioned it to my friends, it would have been in an off-handed remark. I was too scared and embarrassed by my foolishness. Later, I did the same thing in a much more serious way, coming off Utah's

King's Peak, after some 15 straight hours of work, separating from Fred and down-climbing hundreds of feet of cliff face unbelayed on completely unknown terrain. Once again, I was very lucky. Somehow, I have always managed to work my way out of extremely dangerous situations, unsullied, basically unharmed, even if occasionally partially broken.

My parents taught high school science. My father specialized in physics but could also handle astronomy, chemistry, geology, and mathematics and my mom was in biology. My dad did not bring his work home, except for test grading, but my mother prepared nightly lesson plans and stored fruit flies in various locations, which was especially disconcerting because when I was still quite young, I decided that killing animals was unacceptable, and so I became a fanatically strict vegetarian. Thus, I did not like the fact that my mother was incarcerating these little experimental creatures. I'm gratified that she did not work with mice or frogs, at least not at home. During the Depression, my parents had gone to excellent schools (City and Hunter Colleges and Columbia for graduate work). My mother studied with Franz Boaz and Ruth Benedict and worked for Edward Thorndike. My father was a genius who spoke a host of languages and could do anything including practical tasks such as plumbing, wiring, carpentry, photography (shooting, developing, printing), and so on. From the earliest age, I was also exposed to literature, art, music, and all of our books explaining and clarifying physics, chemistry, and radio construction, but also Gilbert and Sullivan, Ibsen, Lynd Ward, *Alice in Wonderland*, and *Thäis* (which my mother mailed to me just a few years ago, when I must have been about 60).

My environment may have been refined and cultured but I was not. Naturally, I listened to classical music both on WNYC and through my parents' collection of 78 RPM recordings, occasionally looked at the large plates in their art books, and picked up the *New York Herald Tribune, Science, Scientific American,* or *Saturday Review,* but mainly I played both indoors and outside. I also traveled in the city alone when I was only six or seven. I would go up to Macy's and wander around as if it were a museum. I never bought anything. I had no money. One day, I somehow managed to find Washington Irving High School. I walked into the office and asked to see my mom. The woman took me to her classroom and knocked. My mom came to the door and looked at the woman at face level; I do not think she saw me at first because I was so tiny. When she looked down, she almost had a stroke. How I managed this is beyond me. And why was I wandering around the city alone, taking trains when I was seven? (My wife insists that if this occurred today, the parents would be arrested.) My adventures are even more astonishing then it might at first appear because my mother was one of the world's great worriers and my father was an extremely strict disciplinarian. But they surprised: When I did not put my toys away when I was two years old (yes, two), my father got rid of them (he was a fanatically meticulous person) but when I was 14 or 15 and announced that I would no longer eat meat, fish, poultry, or seafood, they said fine. When I was just a bit older, instead of complimenting me for staying home and studying, my dad actually said, Go out on a date! And they let me go to the movies alone. I was seven in 1949, when I walked over to The Tribune (near City Hall, so at least 10 blocks away) and saw *The Secret Garden*. The scene in which the young boy rattles his chain so scared me that I walked out, the only time until recently that I ever did that. Seven years later, my dad and I went to this same theater to see *Invasion of the Body Snatchers*. Nevertheless, everything was a struggle and our home was often tense; I got sick.

I went to the Henry Street Settlement House, Manhattan's PS 1 (in Chinatown), and, starting in the first grade, PS 177, which was directly across the street from KV. I was soon walking to school by myself. One day, when I was in the fourth grade, I brought home a St Christopher medallion. My parents were aghast. They apparently feared that I was about to convert to Catholicism. They sent me off to a Yeshiva, not just any orthodox school, but MTJ, the home of one of the world's foremost responsa rabbis, not that I knew anything about such matters when I was engaged in learning to pray. I spent the fourth, fifth, and part of the sixth grades worshipping god—from eight am to after noon every day. A limited amount of time was allocated to learning to read and write Hebrew, but mainly we prayed. Rabbi Rand let me know that when davening the *Shemona Esrei*, one does not move one's feet, even if a snake attacks. Good to know: Scare the children out of their wits. Never once, in all of those years did we study the holy texts or commentaries (analogous to the Vedas in Hinduism or the Bible or Aquinas in Christianity). A friend who stayed on through the eighth grade told me that eventually he did get to do some Torah and Talmud. From one to five we managed a few secular subjects. Of course, I walked the precarious half-mile home alone and at times in the darkening evening. One afternoon, Mr. Levine got angry. He had a child hanging by his hands on a closet hasp and he proceeded to whip him with his belt. I think it may have been his son. (Females did not attend MTJ.) I was shocked, even though I was used to corporal punishment (infractions were punished by smacking a child's hands with a ruler). (I was not afraid; I never feared bullies; and even though I was a diminutive child, I stood up to these creeps, even in high school, when they were causing harm to others; they never bothered me. Somehow I always prevailed.) I went home and announced with absolute certitude that my days at MTJ were at an end. My parents tried to change my mind, rationally and logically, but I refused to capitulate. It was an unpleasant and wasteful religious environment. Next came a visit to our home: The principal, Rabbi Swatitsky, and a lovely instructor, Rabbi Frankel, sat in our living room and importuned me to return (my parents were very active at the school and paid tuition), promising that such an outrage would never recur. I stood my ground. I returned to the equally ineffective 177, finished the sixth grade, and then shifted to Junior High 65, one of the worst schools I have ever encountered: 18 year olds, held back for 10 years, smoked pot, had guns, and trashed my homeroom. I lasted one semester. Every morning, my five year old brother and I would take the Avenue B bus to the East Side Hebrew Institute, where I dropped him off; I then headed back downtown to 65. This was a big responsibility for a 12 year old. We left Manhattan and moved to Staten Island, the most boring of New York's five boroughs. I completed the seventh grade, and off we went to Vermont for the summer. I then managed to change my life.

For my first nine years, we lived in KG5, the K building, G apartment, on the fifth floor. Right next door, in KF5, lived a boy born just 8 months after I was. We began to play together as soon as we had any autonomy and continue to do so today. We raced through the marble halls, ran up and down the staircases, wandered around on the roof, 14 stories above the ground, and did many things we should have abjured. Marvin eventually also attended MTJ. I spent much of my early life in his home where his parents were very kind to me. In 1949 or 1950, a few years after my brother, Don, was born, we moved to a three bedroom apartment in another building. IE7 was magnificent. It had views of both the inner courtyard and South Street. For hours at a stretch, I would sit here up high, looking down watching men construct the South Street Viaduct, which connected the East River Drive with the now defunct West Side Highway. Marvin and I still played handball, biked, wandered around, and had fun, but I also began to hang out with some tough gang kids. We called ourselves the Warriors and we were a minor division of a group of older children. One day, I watched Mickey and another kid bang away at each other with bricks. We did not really inhabit the blackboard jungle and we never rumbled with other gangs. (The only fatality I recall during these years occurred when a lunatic shot someone in the street from high up in an apartment way over on Avenue D near Fourth Street. It was a different time and children were not interested in selling drugs or making money.)

I think my climbing affliction must have manifested itself in these early days on fences, since there were no cliffs or mountains in the city. After the tenements on Cherry Street were demolished, they were replaced by an enormous park whose fence was an impediment to entry, especially if the gate was locked. And this was no ordinary chain link fence that ended at about ten feet; this one went up almost to heaven. Why the city fathers thought it necessary to build such a high barrier is beyond me, but there it was and I climbed up and over with some frequency. A fall would have been extremely deleterious since the ground surface was made of unforgiving concrete. The links were typically wide enough to take the toe of one's footwear and the top was unencumbered with barbed wire (which in any case is no real impediment when climbing at a corner, where the stanchion often foolishly rises above the wire so that one may use it as a handhold). I never fell. I don't think that I have climbed a high chain link fence in 50 years but the tactile memory remains extremely visceral.

In 1945, my uncles returned from their war service. Lou, the marine, came from Okinawa directly to our apartment where he slept. I think that the survival rate for Americans on Okinawa was ten percent. Lou promised that if he survived, he would lead a religious life; he did and he does. The only injury I noticed was a slightly swollen hand. I was three or four but I recall all of this vividly. Hy was back from the army and spent a year in the hospital. My dad would go to Brooklyn to help his sister (Hy's wife) by working in their candy store. Dave came home from the Navy. My dad did not serve because he taught aeronautics to high school students who were able to go through military training more quickly. I once saw the plane that they had deconstructed, carried up into a classroom, and then put back together.

At about the same time, that is when I was about four, I slipped and hit my chin on the kitchen table; it required stitches to staunch the bleeding. All of my many injuries have been similar: minor, self-induced problems that were easily repaired. This was not the case with Marvin, who, while attending MTJ, an ostensibly severe parochial school, allowed Bluval to spin him around (while playing ringolevio); he fell, hit his head, and was on the critical list for three horrible days. Even though I was just ten or so, I remember how hard this was on his mom. Marvin had other really serious injuries. And I stupidly played roughly with Johnny Neumark, who took my head in his two hands and smashed it onto the hard marble hall floor. This broke my front tooth. I am not big on cosmetics so the tooth remains angular.

When one lives in close proximity to some 6,000 people, many unusual things occur. One day, a girl (perhaps a teenager) living across the hall from Marvin, locked herself in her parents' apartment. The front doors at KV were made of steel; the industrial locks, bolts, and chains made the doors impregnable. The fire department came, put up a ladder all the way to the fifth floor, and somehow gained entry. I suppose the young lady got punished. Two girls, Alice and Anne, lived in one of the 12 buildings. I do not remember them, but it is quite possible that they were in my PS 177 classes. When McCarthy (and Hoover) began to wreak havoc with American democracy in order to protect it (does this sound vaguely familiar?), the entire family decamped to France. Many years passed. One day, somehow, my wife's parents met these folks in Paris, and later Anne and her husband visited us in Vermont. More potently, I came to my mom and demanded, "Why aren't you collecting signatures like Punch and Judy's mother?" My mom, who was neurotically polite, countered with, "Mind your own business!" It turned out that if anyone had suspected that my parents sympathized with the Rosenbergs whom they knew (they did not), they would have lost their civil service jobs. Not enough signatures were collected, and the Rosenbergs were executed. Their children, Michael and Robert were adopted by Abel Meeropol, who (naturally) was my wife's dad's teacher! He also wrote the Billie Holiday song, "Strange Fruit." The world is smaller than we think.

I would be remiss if I failed to mention Mazie. She lived at the end of our K building hall with her brother and sister. She always dressed in outlandish clothing and was very heavily made-up. I doubt that I ever saw her without a pill box hat on her head. Her sister was more subdued; her brother was very large, about 400 pounds. Naturally, young children would hassle these people, perhaps bang on their door and then run away or make fun of them, because they were frightened by those who are different. I liked them and got along fine. Many years later, I went to Coney Island and visited Mazie at the bumper car pavilion, which she owned. She let me zip around for nothing. Next door to Mazie lived Mrs. Wassermann; she was ill and her passing, when I was eight or nine, was my first experience with death.

Some of the KV buildings contained stores that sold drugs or sandwiches, haircuts or liquor. In the K building we had three establishments that one could enter from Catherine Street, and it was possible to walk through and come out in the hall near the elevator that led to our apartment. The food and toy stores were always of some interest, but most important to me was Hanscom, a chain bakery that offered various small pastries and danish that I liked very much. I still enjoy good danish more than any other desert, although examples of this are hard to locate even in big cities. In St Cloud or Burlington, the best one can usually do is Entenmann's. The inexpensive apricot, prune, or raspberry danish of my youth, with which I regaled myself on a daily basis, is long gone.

When I was just a little tyke, the dermatological problem I inherited acted up, and there I was in my parents' bed playing with some glitter. The austere and well-known Dr. Kugelmass made a house call. Upon noticing the glitter, he told my mother what he thought of her. He was not very diplomatic; he also, naturally, had nothing tangible to offer. All of the junk that passed for medicine in the mid-1940s had no effect. It was like using hydrogen peroxide, before antibiotics were readily available, to cure a bad infection. I suffered. I had seen enough doctors to know that the only thing that worked were ACTH injections, so when my dad tried to convince me to see Dr. Brown, I balked. I must have been about eight or nine, but strong-willed, stubborn, and, in this case, right. I recall sitting in the car under the Manhattan Bridge in Chinatown while my dad argued rationally and logically for an hour. I countered; I won.

TWO
VERMONT

Don was born on December 26, 1947. I remember that my dad took my mother to the hospital, although I don't know how he accomplished this because New York was in the midst of one of its most magnificent snow storms. I must have stayed at Marvin's house. Six months later, my father became summer camp director at Timberline. He liked this so much that my parents decided to open their own camp. In July of 1949, we bought a 1933 Plymouth and drove to southern Vermont looking for a house and land. We found them in tiny West Wardsboro, up high in the forested mountains. During the twenties and thirties, the Grumbachers, who still own a large artists' supply company, had used this as their summer refuge, but the enormous old house had lain fallow for quite some time. It was a mess: no electricity (except for a 32 volt generating plant replete with 16 large glass-wall batteries); no water; no gas; no oil; no infrastructure. We stayed at a tiny cabin in Wilmington, and every day my dad and I drove to the house and worked. We dealt with room after room; moved furniture around; and tidied, swept, and cleaned. One day, we were working in the darkened basement when a hinged potato bin cover crashed down on my father's hands trapping him. He calmly called me over and asked me to lift it up, which I did. Despite my diminutive size, I have always been extremely strong, able to dead-lift hundreds of pounds especially when doing physical work (haying, stone removal, cement bag delivery), rather than in a gym. By July 1950, the house was habitable. In 1951, Marvin and his family paid us a visit. We picked our blackberries and raspberries and baked pies. We were able to produce some electricity and the fire department flushed the pipes from the five springs located high on the mountain and so we had an unreliable source of water. We continued this process for almost 40 summers: repairing, extending, building, drilling, installing commercial electricity, and eventually eliminating all but one of the outlying buildings: the goose house, shed, ice house, chicken and breeder houses, and even the exquisite old barn all had to go. Some of these were burned in an enormous bonfire, through which we were able to walk, with flames leaping on either side! The house now has 25 rooms and belongs to a neighbor, although I kept most of the land, on which I built our new house with hand tools basically by myself. The camp remained a dream despite the fact that by the second summer my parents had purchased 15 or 20 army surplus cabinets, a commercial stove, and other necessary accoutrements to service the imaginary young campers, and had Allied Van Lines truck them all the way from the City to our new house. As a tiny nine year old, I carried the cabinets off the truck and up the stairs; the driver commented that I would be a very strong adult.

The out-building in the best shape was the enormous barn and it was also the last to go. My dad wanted to improve the view and so he invited some men to deconstruct the barn and take the excellent wood (six by six hand-hewn timbers and 1 by 8 sheathing) away. I do doubt that they paid him anything for this valuable lumber, because he never sold his possessions; he just gave them away. Until this unpleasant day, we used the barn for winter storage: the enormous wooden shutters for the terrace windows, the large swing, the slide, the sandbox, and other accoutrements that my dad and I hauled in both directions every summer. This was hard work even for a strong child. Indeed, the work never ended. The barn had a second story hayloft and a smaller third story on which a very heavy, 19th century buckboard was stored. It lived up there for decades, but one summer we somehow managed to lower it to the ground. My brother and I pushed it around. What we really needed was a horse to pull it. One day, when we were still quite young, a horse pulling a wagon came down from the other direction. We were stunned. We did not know that the road went anywhere since no one had ever passed by before. The only strangers who came up were people, who were staying at the Green Mountain House (hotel), out for a stroll in order to take in the beautiful view. As the years passed, the view disappeared in a forest of fast-growing conifers.

Figure 1: Bob and Don on slide around 1951, by Sigrid Hauptman
Figure 2: Bob in around 1948, by Irving Hauptman

My dad parked the Plymouth (square with a soft, tarred roof and a floor shift) uphill. One day it would not start. We could have walked the half mile down to the Newell farm and borrowed a battery (we did not have a phone) or Tom probably would have driven up and used jumper cables to start it. But I am fairly certain that this never occurred to my father who was very self-sufficient. Instead, what followed turned out to be the most astounding achievement of our lives, at least according to my mother. My dad and his nine or ten year old son (that would be me) turned the car around. This was difficult to do on a hill and also rather dangerous because with both of us pushing, the vehicle could have taken off downhill and hurt itself or us. (Something similar happened many years later, when I was sugaring deep in the Newell woods. The driver had parked the double-tracked bulldozer—used to pull the sap collection tank, as horses had in the past—downhill. We were eating lunch and all of a sudden the dozer took off and crashed into a tree. I still shudder when I think that one of us might have been crossing in front of it, when it decided to exercise its freedom.) We then pushed the car; it picked up speed on the descent and started. Today, cars have transmission interlocks and will not start when rolling; even those with manual transmissions are so encumbered.

The exquisite house is about 250 years old, porous, and often unheated. While New York and Boston swelter during the summer, the Vermont woods at almost 2,000 feet get cold at night. One evening, when I was in Oklahoma, suffering in the hundred degree heat, my dad concluded a phone conversation by saying, I have to go and build a fire for your mother. In the early days she would stand over the kitchen stove burner rubbing her hands together, since it could be 40 degrees in the mornings. A fireplace and space heaters alleviated matters. Eventually, my dad had a hot water heating system installed throughout the house but he did not live to see it in operation. As the years have passed, the average summer temperature has increased dramatically so that most recently, during the summer months, we have not needed any heat at all.

At the end of one early summer, I asked my parents if I could return to the city a few days before they were to close up the house and leave for the new school year. Astonishingly, they said yes, so I took a bus or train to New York and managed to get to KV, where I entered our apartment and turned on the fuses. I then invited some friends over and we had pasta for dinner. None of this is very exciting except that I was only eleven years old. This is another example of how my excitable mother and strict father acted uncharacteristically, and, according to Terry, you will recall, would, today, be brought before a judge for child abuse, although it is good to keep in mind the fact that things were very different 60 years ago: Murder was harshly punished, rape was infrequently reported, and kidnapping was extremely rare. (Today, rape and kidnapping are ubiquitous, and the perpetrators often go unpunished.)

I spent my school days dreaming of our return to the woods and by the sixth summer, I was so enamored of the rural life that I did not want to leave. I was now 13 and had developed into an extremely independent thinker. I came up with the possibility of staying on alone. One might have thought that my excitable mother and my strict father would have rejected my request out of hand, but again I was much surprised to find that they agreed. My father asked around and someone suggested that we stop at a house on the outskirts of Brattleboro and make inquiry concerning the possibility of my boarding with the family for the school year. We knocked; they answered and said no thanks. We tried once more in South Londonderry and astonishingly they said yes. A few weeks later, my parents dropped me off at David Melendy's farm house, where I lived from September 1954 until June 1955, when I graduated as the eighth grade valedictorian; my speech concerned the sands of time. At one point during the 1954 summer, our neighbor, Tom, had asked my dad if he could have our sugar house, the only building that lay far from our home, deep in the woods in close proximity to a sugarbush that no longer existed, because the previous owner had logged the land; he had cut a quarter of a million board feet of lumber and did not care to preserve the maple trees that comprised the sugarbush. My dad said yes, and Tom deconstructed it and rebuilt it about a mile away on his land. Tom presented my father with $50, which he refused to take; they may have gone around a number of times, but the upshot was that when I arrived at the

Melendys, I had $50 in my pocket.

My parents made a negligible payment, something like seven dollars a month. Along with Harold, who was also in the eighth grade, I worked very hard feeding the goats, horse, and chickens; milking; digging potatoes; cutting wood; stoking the furnace; and on weekends, going into the forest to cut both firewood and pulp. On one occasion, just Mr. Melendy and I went up high on what is now Magic Mountain, the ski area, to work. He parked the pickup and told me to stay in the cab. He went over to a very large tree, planted some dynamite, and set it off. (This is a rather unusual form of logging!) One very cold morning, I went out to the barn to milk the goats and I discovered a newborn kid. I ran back to the house and we then brought the baby goat into the dining room. Mr. Melendy put up some wire mesh to contain it; we fed it milk and it prospered, although its hooves slipped on the linoleum floor. It would not have survived in the cold barn. In fact, one of the three adults (Grizelda, Tarzan, and an anonymous nanny) did succumb.

My dad visited me at both Christmas and Easter, since as a teacher, he had vacations, but I did not see my mother or brother during this entire 10 month span, and this was very difficult, not because I was lonely, but because my mom was deathly ill. Her colitis had reached a point where she could no longer derive nutrients from food, and had she not opted for one of the first colostomies, she would have died. As it happened, the operation took place just as I returned to New York, and so I went alone to Mt Sinai. Things were, once again, very different then: Children were not allowed in the hospital. I snuck into a stairwell, walked up nine flights, and did get to see her. She came home and lived almost 50 more years. In those early days, she would go around to hospitals and encourage depressed people by confirming that one can lead a perfectly normal life—even swimming—despite this seriously unpleasant surgery.

I did well with my peers, though they were as different from me as an Australian Aborigine is from a Texan. They never hassled, harmed, bullied, or threatened; they never made fun of me or my city ways, my accent, or vocabulary, my unusual religion, or my diminutive physique. I already knew how to drive, saw, cut, chop, feed, and other practical farm necessities. What I did not know I learned. I helped, and the children were pleasant. One day, we somehow got an old car to run and Harold and Calvin took turns driving in a field; then they let me drive too! If they did bear me any animosity, they never showed it. And it was not as if they were angels who adored each other. Part way through the year, a few of us were in the sixth/seventh/eighth grade classroom alone. The wonderful teacher and principal, Mrs Stocker, was elsewhere. For some reason, the otherwise fairly placid Calvin and Butch were fighting, really trying to hurt each other. I did not participate in such serious bouts, ever, though I certainly have fought toughly for fun whether in roughhouse wrestling (my front tooth is still broken) or serious (and dangerous) karate (kempo) training. We belonged to the Shirts and Skirts, a square dancing group. I have never liked any type of dancing and so I was not very good. Thus, when we drove all the way to Tufts in Boston to perform at what was probably a competition, I did not participate; I was undoubtedly relieved.

This was one of the seminal years of my life. Lamentably, of the five Melendy children, David (Jr.) was in Colorado and I never met him; Margaret passed away many years ago; and Harold, whom I never saw again, more recently. I did visit with Alice around 2000. I found out where she lived and knocked on her door. I asked if she were Alice Melendy (the 11 year old girl with whom I had lived for a year); she replied affirmatively. I then wondered whether she knew who this 60 year old man was. She said yes, and impossibly, she did.

At that time, most Vermont high school graduates did not consider college (the many currently available technical and community colleges did not even exist). The only thing on the agenda was work, in manufacturing, farming, logging, clerking, or driving; the tourist industry, especially skiing, had not yet come into its own. Options were limited. Another possibility, especially for woman who planned to stay home, was immediate marriage. The fifth Melendy child, Clara, graduated from high school on Friday and married on Sunday. She must have been 17 or 18 but was a real adult to my 13 year old child. I had almost nothing to do with her during the course of that entire year, although she lived, ate, and slept in the same small house. She was treated differently and had her parents passed away she could easily have assumed responsibility and taken over as a surrogate and effective parent. To a lesser extent, naturally, in rural areas (and not only in Vermont), this is still pretty much how matters unfold today.

During this year, Harold and I did two extremely foolish things; they both involved fire and they both could have resulted in real tragedies. We were only 13, but knew better when we built a little enclosure out of bales in the hayloft and smoked a cigarette or two. Nothing happened. This was not the case when we were doing an assigned task, which was to use an old fashioned blowtorch, the type with an open ignition trough for kerosene at the top. As we thawed the basement pipes, a stray spark ignited the wood chips on the floor and the fire immediately shot up along the wooden walls. We managed to extinguish all of this and no one ever found out. The fire was not our fault but still we were responsible. The Melendys had already lost one home to fire. I am fairly certain that they would not have dealt well with another debacle.

Since 1949, I have spent all or at least part of every summer on our West Wardsboro property. After I married Terry in 1968, she joined me, and now so too does Kira, our eleven year old daughter. During my youth, I naturally stayed with my parents; in 1972, I retrofitted the last remaining out-building, a blacksmith shop. Terry and I lived in this rough, bathroomless cabin year-round until 1977, when we went to Pittsburgh. I worked at a hardware store, sugared (made maple syrup), hayed, painted roofs, but mainly read, sometimes 18 hours a day. For two years I also studied for a second Master's degree by commuting to SUNY-Albany once or twice a week. Despite the very harsh storms and our old two-wheel drive vehicle, I only missed one day. This was another idyllic period: During the winters, the road was unplowed so that we had to walk or ski in. Visitors had to do the same thing and so only the ambitious visited. Sometimes, I would ski the mile down to town to get the paper and mail and then climb back up. We never got cabin fever and we got along as well as we did in much larger quarters, which doesn't mean we agreed on theoretical, academic, or practical matters.

At some point during this five year period, we met our neighbor, Robert Penn Warren, his wife Eleanor Clark, and their children, Rosanna and Gabriel, who were quite young. Red, as he was called, was still teaching at Yale, but the family would come up for holidays and during the summer. The Warrens spent the 1975-1976 academic year in Italy. (I recall mailing them Lionel Trilling's obituary, because I thought *The New York Times* might not make it to Florence.) When they returned, Eleanor had lost her sight and could no longer drive, although she sometimes tried. One extraordinary day, I drove her in her Mercedes to New Haven to buy extremely large pads and markers so that she could continue to write. She did and produced *Eyes, etc.* (in which Terry and I make an appearance). While she was rummaging around in the art store, she arranged for Donald Gallup, the well-known Eliot and Pound bibliographer, to give me a private tour of Yale's Beinecke library. He took me to some wonderful places that normal patrons do not get to visit: In the bowels of the library, I discovered hundreds of shelves and filing cabinets containing original manuscripts from America's great writers. One drawer was sealed and covered with a note stating, Do not open until 2020: Alice B. Toklas.

During the six hours we spent in the car, Eleanor regaled me with many wonderful stories. I very foolishly asked about her visual impairment; she explained much too precisely, and I felt the blood rush out of my head. I de-accelerated from 70 down to zero as I pulled off to the side of Interstate 91 to recuperate. She wondered what I was doing. I explained that I could not drive the car if I fainted. Eleanor was an extremely tough Yankee and would not allow a major visual impairment to stop her from driving on an Interstate, and so she casually but forcefully asked whether I wanted her to drive! I declined her offer and soon recovered and we continued on our way.

Over the years, we were gratified to be invited to their home many times, to eat a formal dinner (they always had vegetables for the vegetarians) and discuss literature and the twentieth century's great writers and artists, many of whom the Warrens had known. Indeed, Katherine Anne Porter is Rosanna's godmother and Max Ernst her godfather, I think. But the single greatest honor they bestowed was to invite us to the premiere of Ken Burns's Huey Long film. It took place at their home and only seven people were in attendance: Burns, the Warrens, Sidney and Anne Hook, and Terry and me. By the way, the little town of Wardsboro with the adjacent Stratton on one side and Newfane on the other has been home to an extraordinary array of well-known writers, scholars, and celebrities: the Warrens, the Hooks, Ernest Nagel, John Kenneth Galbraith (who once kindly replied to my letter), Henry Steele Commager, Eugene Rabinowitch (the first editor of *The Bulletin of the Atomic Scientists*), E. G. Marshall (whom my father interrupted as he filmed in New York, and so the scene had to be reshot), and Ron Howard. Now Rosanna owns the house and we try to get together at least once or twice each summer.

At about the same time that I decided to stay in Vermont for the eighth grade, and then to become a vegetarian, I thought that I would like to build a house. I played with this idea for many years, even designing a round, limestone version. By 1982, I realized that a real if humble dwelling was preferable to an imaginary castle, so I laid-out and cleared some land and a 500 foot curved, ascending driveway; in June of 1984, I had a man bulldoze the site and road and dig a foundation hole. A contractor came in and within 24 hours he was gone; he left a substantial concrete foundation for a large if symmetrical house and garage; the entire business measures 25 by 48 feet. I had decided to build the house myself using hand tools (although I did occasionally employ an electric sander and screwdriver, and I did use chainsaws to do the clearing, in which I was helped by a friend for a few hours). This same man, Bob, came to work with me one day a week for ten weeks. We would labor extremely hard and accomplish a lot, and then I would continue until he returned. By the end of the summer, we had completed the entire exterior of the house except for the 18 foot garage door installation and the exterior finish wood. I continued during my many following four-month summers doing almost everything alone: insulation (with help), sheetrock, hardwood floors (back-breaking), all electrical work for two systems (AC and DC), plumbing (threaded galvanized pipe rather than sweated copper and with help with sewerage), masonry (except for the final brickwork above the roofline, because my joints were not very pretty). Years later, I had crushed rock shot onto the basement and garage floors followed by concrete, which we smoothed out. As time passed, I came up with new projects: an additional room hooked on to the back of the house and a deck on both of which Bob also helped; reshingling and a new deck (with help); and a guest room inside the enormous garage (with a bit of help). I had the little dug spring replaced with a drilled, semi-artesian well that supplies all of the water we require and, finally, I was able to have commercial electricity installed despite the exorbitant cost. Until this wonderful day arrived, I maintained my own 3,000 foot telephone line through the woods and produced electricity with a 4,000 watt generator (for lighting, cooking, refrigeration, and hot water) that simultaneously charged a marine battery. This provided 12 volt incandescent light in most of the 10 (now 13) rooms when the generator was silent. We moved in in 1988, and ever since we have been there every summer and during other seasons when sabbaticals allowed.

Building a house requires many skills and one must do a multiplicity of tasks. Different people prefer and are good at different things. Tarring the foundation is unpleasant work for almost anyone and I would guess that most people would enjoy putting up siding. I especially enjoy framing. Here one installs very large structural members (2 x 4s through 2 x12s, in two inch increments), which are used in various installations. A 16 foot 2 x 12 is a heavy member (and weightier if pressure-treated) but I much prefer working with these enormous boards to taping sheetrock or fitting molding into complex configurations. Because I use hand tools, every task took more energy and often more time. I lapped all joists on a central beam, rather than butting them against each other (because this offers additional structural integrity), and I used precut 2 x 6 studs so no cutting was involved in these tasks, but I did have to cut each end of the rafters. Since these roof supports are angular, each 2 x 8 required a ten inch cut. There were 104 of these, 103 of which I did by hand. (I did not use birdmouths, which I distrust.) I also cut every piece of plywood wall and roof sheathing by hand. I enjoyed the work. I happened to be away when the shingles arrived and my dad helped the driver unload almost 20 squares; that's a lot of heavy bundles. Bob and I attached them to the roof.

Electrical work is precise but easy. I did it twice because I wanted to have both AC and DC (12 volts) available in the entire house, so I ran two systems, the first of which requires multiple lines. I used 12/2 Romex for the entire business. I regret it. If I was to build another house, I would use BX (wire encased in metal sheathing, which helpfully also provides an automatic ground) or put all of the plastic coated Romex in conduit, something normally done only in large commercial installations. I dislike the fact that my wiring exists at the mercy of weather incursions, rodents, and sharp objects.

The most unusual aspect is the plumbing. Because I have never worked with sweated copper (although I am familiar with cast iron/oakum/liquid lead), I rented a cutter and die set, and cut galvanized pipe to the required lengths, threaded the ends, and put the pieces together like a puzzle. It gets very cold in Vermont and I wanted to be able to drain the system quickly and efficiently at the conclusion of our summer vacation, so I placed the kitchen and bathroom on opposite sides of a wall. Thus, all of the piping in the basement and above are in a single location. Even the second floor bathroom, which I installed much later, is located nearby. I can clear the system in just a few minutes, pour anti-freeze into the traps, and leave while the hot water tank continues to drain itself. Only occasionally have I had a problem with freezing.

All interior work followed in its appointed fashion. The hardest task was installing ceiling sheetrock by myself. With the correct professional devices, it is simple, but using home-made supports is torture. Once, I asked Terry to hold a partial sheet in place; she dropped it on my head. I continued alone. I built the chimney after everything else was in place. This entailed cutting openings in the floors and roof. For me, this is very trying, because I am destroying something that I just completed, but somehow I managed. The chimney drew nicely, when I connected a lovely, airtight emergency wood burner to it. I subsequently removed this cumbersome object from the living room and installed an emergency gas heater. Some of the rooms are equipped with efficient if expensive electrical baseboard units. Wiring 220 volt devices including a thermostat in the configuration is complex, but I figured it out. Naturally, we have used the gas heater even when no emergency presents itself because it provides immediate warmth; the baseboard units take their time warming things up. I merely mention some of the hundreds of other small and large tasks required to complete a modern dwelling; these include (in my unusual case) exterior and interior foundation drainage systems, buried in earth and concrete, interior walls, 14 large built-in book cases, stairs, windows, doors, kitchen cabinets (twice), phone lines, painting, and so on. We also now have a hot air furnace.

 I retired in 2005 and in 2006 we began to spend the academic year in Colchester and subsequently South Burlington, Vermont. For six years, I suffered because of the toxins (such as formaldehyde) released by the building materials used in new commercial dwellings. Thus, in June of 2012, we decided to move to our mountain home. We got rid of a lot of toys, clothing, and toxic furniture and then somehow managed to squeeze the rest of our possessions into an already crowded though fairly large house. The first thing I did was to build new book cases to accommodate some of the 3,000 books we brought with us as well as an additional large linen cabinet for towels and quilts of which we have perhaps 50 including many extremely beautiful knitted Afghans that Terry's grandmother made for us. She was still creating these when she was close to 100. It took many months to put away all of the books, tens of thousands of papers, and other stuff contained in the 200 or so boxes that we brought with us. In addition, Terry's countless and often enormous paintings required organizing and storing. It will therefore come as no surprise that I decided, after much prodding, to build a small addition, the first floor of which is Terry's new studio and above it, under the steeply slanted roof, is a new room for Kira, who has been sleeping in an old guest room.

 The house is deep in the woods surrounded by hundreds of thousands of acres of forested land. Even today, after years of development and the addition of new houses along this formerly empty country lane, there are very few people around. They build their enormous and expensive houses (including the largest private log home in the US, put together in Montana, deconstructed, shipped to Vermont on three flatbed trailers, and reconstructed on land contiguous with ours), but spend almost no time away from their distant primary residences where they must be in order to earn a living as an Interstate highway contractor, firefighter, computer software executive, teacher, minister, or radiologist. We are alone with the animals. Kira, in just a few years, has already managed to see most Vermont mammals (bear and cub, deer, moose, fox, rabbit, raccoon, beaver, skunk, and even a fisher, which I have spotted only once in more than 60 years spent wandering around sometimes deep in the silent and distant woods). I too have seen most of these as well as others (porcupine, weasel) and many species of identified birds, which

Figure 3: My Parents in Front of Their Vermont House, 1985
Figure 4: The House I Built, 1993

I search for diligently, racing through the woods following a barn owl or standing in an open field mesmerized by an enormous pileated woodpecker, but even I have never seen a river otter, bobcat, or mountain lion! So, all and all, this is a small piece of paradise, but I recently had an astonishing revelation: As much as I care about the house that I built, I prefer our nearby old home more, perhaps because I invested a part of my life and soul in its reconstruction. My brother, by the way, never liked being in the country at all.

In the spring of 2012, because I was extremely allergic to something in our Burlington area apartments (where I had no physical responsibilities), we moved to our property, where, for the first time in almost 35 years, we planned to stay permanently. I had forgotten how much harder life in the wintery woods can be, and especially for me, since I do everything myself. No one picks up our vast quantities of trash and recyclables. No one removes fallen trees. No one arrives after a two foot snowfall to plow the 500 foot driveway. I must rake the enormous and complex roof to avoid collapse but especially to prevent ice dams which cause leaks; I must shovel pathways; I must clear the car (which is sometimes parked at the bottom of the drive) of snow and ice; and I must use a snow blower to clear the road. If I do not, we can of course walk down to the parking spot, which is only a tenth of a mile away, but then the oil and gas trucks cannot deliver fuel. I seem to think that I am much too old for all of this; despite the heat and hurricanes, alligators and pythons, Florida is calling me, at least for the four really hard months. Naturally, just when we plan to return, in late March or early April, we will have to contend with mud season, which, when bad, makes the dirt roads impassable. We will have to carry our cats in on our shoulders.

THREE
EDUCATION

I am perhaps the only person to have attended a Jewish Yeshiva, a Lutheran college, a Catholic university, a Protestant theological seminary, and a zen-do. I wasn't searching for anything. I had already found it: I have been a committed atheist since I was 14 years old. I prayed; I earned a degree at Wagner College; I studied philosophy and art history but also managed something most unusual at the University of Innsbruck: I crossed faculty lines, that is, I took classes in both the humanistic branch of the university but also sat in with the cassocked theology students for "Die Kunst der Katakomben," taught by a peripatetic Jesuit just back from his Roman underground observations. I worked on esoteric languages at the Pittsburgh Theological Seminary where one may sit at Karl Barth's desk and also study advanced Hebrew Grammar and Middle-Kingdom Egyptian (in hieroglyphic script; hieratic is too difficult; demotic impossible). In high school, I uncaringly studied a broad diversity of subjects including metal machining, chemistry, and law. After spending time in the work force, I realized that I wanted to learn and in a formal way and so I went to college with very serious intentions. I studied everything and with scrupulous care devoting my time, for the most part, to school work rather than earning extra money, partying, drinking, dating, or socializing. By the time I completed all of my graduate work 15 years later, I had managed to formally take at least one class in virtually every general disciplinary area (with the exception of anthropology, business, and medicine): computer science, ten languages (e.g., Russian, Arabic, Anglo-Saxon, Icelandic), sociology of science, economics, physics, modern algebra, psychology, literature, applied music (piano). I left nothing out. I audited many classes for pleasure (and continued to do so even as a full professor), but apparently took enough credit-generating courses to earn a BA (cum laude), two Master's, a doctorate, and a second doctorate (ABD) for which I was given a certificate of advanced study. I like formal education—both as a student and as an instructor.

I did not go directly to college. When I was in high school, all I cared about was skiing, which is a very expensive pursuit especially if one lives in New York City and has to travel to Vermont on weekends. I worked and earned a paltry amount of money and devoted it exclusively to equipment, travel, and lift tickets, but a part-time library job at 95 cents an hour made life difficult. As a senior, I worked full-time every other week, but the $33 I managed every ten days was still inadequate. Thus, when I graduated, I decided to work for a year, buy a car, and ski. I had an unpleasant, tedious job at Helbros Watch Company, which was located at the corner of Fifth Avenue and 48th Street, right next to Brentanos book store. I learned very little about watch mechanics and came to dislike my primary task, which was to open and label the 600 items that arrived every day for repair: "I wore my water-resistant watch in the shower and now it no longer works." Perhaps you rusted the pendulum, delubricated the bearings, or welded the bezel to the case! You should have thought about the meaning of the word resistant in the context of a cheap timepiece. Helbros was a far cry from Rolex. I used every excuse to stay home. I soon realized that in order to survive, I would have to go to college and so I took the SATs. I do not believe that I spent any time studying the many available guides or re-practicing geometric computations or honing my spelling skills. I just took the test and did okay despite my horrible high school performance. Now, with an additional 80 or so points added to my 1960 score (the tests were apparently more challenging then), I must be in the 99th percentile (I am exaggerating). I applied to one school, Wagner College, which is located just a few miles from where we lived. I arrived in September of 1960 and began with five difficult classes: world literature, German, modern algebra, history, and religion. The latter was a requirement. I applied myself with real committed diligence and did well. Grade inflation came along much later as one can see from my position in the hierarchy of graduates. I had a 3.2 GPA, but I was right at the top. Only one male graduated summa cum laude and one female magna cum laude. Out of 750 graduating students, I was in third place (cum laude).

My second year was similar. I read, I studied, I learned: I got perfect grades (in German, Russian, and two other subjects). At some point, my mother informed me that the college was planning to inaugurate a new program abroad for my junior year. I was skeptical at first, but then realized that it was a good idea because I could speak German and ski in the Alps. I did not consider travel opportunities or cultural enhancements in art, architecture, music, opera, or theater. I soon amended my attitude. A group of about 50 students from all over the country left New York in September of 1962. We sailed first to Boston and then across the Atlantic on the Queen Frederica, a desolate, old Greek liner that seemed to plow through the water at two knots an hour. Fourteen days later, we arrived in Naples. (A friend once flew from New York to Paris in three hours!) I spent many of these days lying in my bunk because the roiling sea made me ill. I used the time efficaciously memorizing poems including "Jabberwocky" in its entirety. The Austrian buses that were to meet us were there and we spent some days moving up through Italy visiting an American military cemetery, Capri, and cultural and tourist attractions in Rome, Florence, and Venice. Everything went smoothly, and Dr. and Mrs. Pinette maintained superb order, which does not mean that my peers did not drink or smoke or eat pepperoni pizza, did not rock or roll and have fun. We really were an exceptional group of good youngsters and nothing untoward or scary or horrible occurred.

For the next nine months, we lived in the Hotel Weisses Kreuz on Römerstrasse in Bregenz, the capital of Voralberg, the westernmost province in Austria. Hochdeutsch, the language of the cultured, academy, and broadcasting industry, could always be used and understood, but this area's dialect is Alemanisch (similar to Bavarian) and it is quite simply incomprehensible (*Ich bin draussen gewesen* comes out *i bin ousiksee*). I stuck with normal German and took classes at the Wagner Institute. Dr. Pinette, fluent in four languages, directed the program and taught. When one of the local instructors turned out to be deficient in his knowledge of art history, Dr. Pinette took over and did an outstanding job. I still recall challenging him in class when he insisted that Bach was the greatest classical composer. I perhaps preferred Vivaldi (who purportedly wrote one concerto 600 times) and may have foolishly cited him as a possible competitor. Dr. Pinette implied that I lacked musical maturity; perhaps he was right. I also took a Russian refresher course; the instructor spoke German and Russian but not English. I would sometimes translate his remarks from German into English (as I had done when I was a child at MTJ, trying to bring the instructor's Yiddish into English despite the fact that my own comprehension was rather limited).

The most powerful lesson I learned while living in Europe is that the Second World War had a very different and overwhelming effect on those who experienced it firsthand and this is so even for Americans who lost loved ones who had served in the armed forces. This is obviously because the war was fought in these people's cities, towns, and fields, along their streets and paths; the people often lacked basic human necessities and at times did not have anything to eat. (Some years later in America, I met a young woman who, as a child in Holland, was fed tulip bulbs, which resulted in ulcerative colitis, a lifelong affliction.) Whenever these adults have to indicate something in chronological sequence, they say, *Before the War, I lived in Salzburg* or, *After the war, I earned more money*. This unnecessary and wasteful conflagration was, and probably still is, a major physical, mental, emotional, and spiritual dividing line in these people's lives. Older Americans ignore it (those who participated in the fighting often refuse to discuss their experiences); for younger folks, the First and Second World Wars are as alien as those fought by the Spartans or Persians so long ago.

I additionally attended the local classical *Gymnasium*, where students studied every subject including Greek and Latin. These were the lucky few who had the opportunity to go on to the university. Others, who at the age of 12 had opted to apprentice, were stuck in their waitering or catering or plumbing jobs for eternity. It was here that I met Seppel Manhart, who, later, given the opportunity to climb Kilimanjaro or spend a year in America, chose an American Field Service Exchange student program. He already spoke English, but when in another land, I always limit myself to its language as much as possible, since the ostensible reason for my residence is to learn to speak Italian or Icelandic or German. I refused to speak English except with monolingual fellow Wagner students.

This year changed my life, and a high percentage of my peers would offer the same assessment. We learned about art and history and culture. We visited many countries and skied and climbed. We played basketball and soccer. (We were ticketed for playing in the street and for doing the twist, which was illegal.) We all caught the travel lure, and to this day many of us live abroad or travel constantly to the most bizarre places. One lives in Laos, one in England, one spent time in Thailand, and one flew professionally to Haiti and then went to Southeast Asia over and over again. I managed to travel, visit, and live abroad frequently. Long after we married, Terry and I circumnavigated the globe stopping in wonderful places like Fiji, Australia, New Zealand, Japan, Hong Kong, China, England, and Israel. Our ticket allowed for many additional locations (Singapore, Thailand, Italy) but various impediments impinged. I have been in all 50 states, many Canadian provinces, and some 40 countries in most parts of the world. Before studying in Bregenz, I had never been south or west of New Jersey. New York City and Vermont bounded my world.

Figure 5: Bob in Hawaii, by Terry Hauptman

I returned to Wagner for my senior year. Naturally, I had put off two semesters of science and one of psychology as long as I could, but now it was time. I got a C in the first semester of physics despite the fact that I had no distractions. Then I looked ahead: I knew that I would return to Europe for my first year of graduate study and I needed to earn some money. My parents had paid for college but I now had to assume financial independence. I undertook a full time job with the post office. I worked from five in the evening until one AM at which time I was often adjured to work for two more hours. One could not refuse the government! And then someone requested an additional two hours but even the US government could not force me to work 12 straight hours. (When studying for exams, a note would eliminate the overtime entirely.) I got plenty of sleep even though the trip back to Staten Island (by subway, ferry, and bus) took an inordinate amount of time). I never missed classes and did excellent work including a lovely presentation on German painting with slides borrowed from the Metropolitan Museum of Art. But something took a toll because once again I managed only a C in physics and another in psychology (for a total of six during the course of my entire 15 year higher education student career). Bruce, who had been in Bregenz, and I left for the University of Innsbruck in the fall of 1964.

This was the best year of my life. I read, I spoke German, I attended classes (20 hours a week) in literature, philosophy, and especially art history. Teutonic lecturers tend to be boring, often merely reading their notes or books to those who actually attend class (which is not required in many cases). Everything is very formal; I once heard an instructor address a close colleague as Herr Dr. Lutorotti, I think. Dr. Windischer, the philosophy professor, was different: He spoke well and amusingly, often going off on a Brucknerian tangent. He was so revered that it was difficult to find a seat in the large lecture hall, and his entrance was always greeted with raucous foot stomping (a mark of real affection). I studied *Philosophie der Gegenwart* and also took a seminar on Karl Jaspers with him. I skied, I climbed, and I traveled as far as Turkey and Israel, Norway and Sweden, Spain and Portugal—sometimes on the Orient Express, sometimes on ocean-going ferries, sometimes hitchhiking. I survived Jordanian bandits, psychotic Portuguese drivers, mad German mothers, and a lack of funds. Bruce became disenchanted: He wanted to work on an American degree and so at the end of the first semester, he returned to New Jersey where he eventually earned a doctorate in German at Rutgers University. After he left, and during a very quiet time at school, I was walking on Maria Theresa Strasse and who should I see but Seppel. We talked and it turned out that he was studying law at the university. He became my roommate at the Internationales Studentenhaus, a new, luxurious, heated, hot-watered dormitory for both men and women from all over the world. We cooked our own meals, were motivated, worked hard, and went skiing between classes. I must have been overly ambitious because a few times a week, I walked along the Inn all way to the Institut Français in order to study French. My world broadened. One day, I was in the dormitory lounge listening to two young women speak perfect English but with the most bizarre accent. I deduced from their vocabulary that they were native speakers but I had no idea which country produced such sounds (although I had been to Southampton where even the locals cannot understand a heavy Cockney accent). It turned out that they had come all the way from South Africa to impress me. Many years later, I discovered that people in the Australian Outback are sometimes incomprehensible (much worse than what you hear in Fosters commercials) but New Zealanders are quite mellifluous. (I like both countries very much, though the former is too hot and dry except when it floods; the latter, closer to the Antarctic, is wonderfully cool and snowy).

As the year progressed, I realized that if I were to get into an American graduate school the next fall, I would have to apply. I filled out forms and arranged for letters, scores, and transcripts and then sent a large packet home. My mom mailed the materials to the various institutions for me. I was accepted for English studies at the University of Vermont, which offered me a $1,000 stipend and Ohio University whose generous $2200 seemed more reasonable. I headed southwest in my newly purchased (used) Alfa Romeo Giulietta Spider. It did not hold much, but I managed to get my clothing, music devices, and study materials into the diminutive trunk. The 500 miles from New York to Athens flew by faster than it did on the innumerable roundtrips that I made during the next six years. Although the housing market there was very tight, I located an excellent apartment and proceeded to study English literature. I taught freshman composition and covered all of my costs. One day, there was a knock on my door. Modrus entered and asked if I would care to have a roommate. He would share the rent and install a phone. Since there was a spare bedroom, I acquiesced. I now had a bit more money to spend on books. A year and a half later, I had my first Master's degree. I continued with classes in comparative literature until May of 1967 and then took a year's leave to live in Greenwich Village, earn some money, and return to Europe to polish my Italian. The English department was flourishing with a host of professors including Donald Stone, Oscar Cargill, Jack Matthews, Daniel Keyes, who had written *Flowers for Algernon*, Walter Tevis, who had written *The Hustler*, and many outstanding instructors including Eric Thompson (theory) and Robert Roe (history of the language) and perhaps 100 graduate students, many with teaching stipends and a few on National Defense Education Act fellowships, which did not require any service to the institution. Here I met many Master's and doctoral students who went on to become quite well-known, William Heyen, Stanley Plumley, and Jim Ragan among them. In the fall, I studied Italian and art history in Florence and the following September, Terry and I were married and I returned to OU and continued my studies. Three years later, with a defense of my dissertation completed, I was awarded a PhD in comparative literature. We left immediately for Iceland.

Studying at the University of Iceland was my reward for all those years of hard work. Most people would have preferred the beach at Waikiki, but I enjoy brutally cold and snowy weather and formal instruction. I worked on modern Icelandic in six hour blocks. The class comprised people from many countries (e.g., East Germany, Sweden, Japan, the US) and the primary instructor was Helga Kress, Bruno Kress's daughter. These folks were so important that, like Vladimir Ashkenazi, they were given nominal dispensations: They did not have to restructure their names in the Icelandic fashion, where one is the son or *dottir* of the father. Thus, I would be Robert Irvingsson and my imaginary sister is Austa Irvingsdottir. Almost all of us had the same goal, to learn to read the sagas in medieval Icelandic, but since no course was offered, we studied the modern version, which, unlike virtually all other tongues, is basically identical to the language of the distant past, in this case, the one that Njal or Erik the Red or Leif Eriksson spoke. It remains unchanged: All of the early conjugations and declensions that hamper Old English or Old High German continue to perplex those who wish to master this complex language. Terry and I moved on to Europe and then returned to the US.

In 1975, I began to study for a second Master's. I commuted from Vermont to SUNY-Albany, worked in the rare book collection, and taught a class in paleography. I entertained myself by playing basketball and briefly studying Quiche, a Mayan language, until the well-known instructor left to help out in Guatemala after the terrible 1976 earthquake struck. Upon completion of the degree two years later, we moved to Pittsburgh, where I began work on a second doctorate. I joyfully completed my courses and then returned to Vermont in order to work on the dissertation, but I became distracted with other scholarly projects and never wrote it.

As a career academic at different institutions, I continued to take and audit classes in many areas including Russian, Italian, history of science, ethics, television production, and photography. I also studied at MIT. Just a few years ago, I applied to Indiana University (and was accepted) in order to earn a doctorate in the history of science, but the cost (an unpaid leave and tuition) was prohibitive. I have long thought about going to law school.

FOUR
INTERESTS

As a young child, I had no abiding hobbies. During the Vermont summers, I began to do carpentry and one day my dad allowed me to cover the naked third floor joists above an enormous 25 by 40 foot room. I used randomly sized boards some of which may have been removed from the outbuildings we deconstructed. The ice house had been yellow, the shed gray. The whole thing was a big, colorful puzzle, getting the variously wide boards to fit together. I cut and nailed. I worked alone. I recall asking my dad for some advice and he said, Work it out. I was 10, maybe 11. Some years later, after returning to New York from my eighth grade Vermont sojourn, I asked if I could build a small room in the open unfinished attic. This was a strange request because placing a rough room in the middle of an upper floor is most unusual especially in a rented apartment. But I did it and alone. I ran 2 x 4s to the high rafters, covered them with panels of masonite, built in both a floor to ceiling closet with a secret hiding place for my excellent if proscribed comic collection (Donald Duck, Tarzan, Captain Marvel, and a few others), and a television set (the screen flush with the wall). I installed a microswitch on the stairwell door; when someone opened it, a light flashed. It was cold in the winter and hot in the late spring and early fall. I liked being alone. I coped. I was 14 years old. I spent some of my four high school years up in this aerie, although I did my homework at my dad's living room desk while listening to rock and roll evolve.

I watched American Bandstand and listened to the top 40 songs on the radio every day. From the Penguins, Elvis, and Little Richard to Shirley and Lee and Fats Domino, this period, which coincided with my high school career, is the best in rock n roll; the years 1959 through 1964 also produced some excellent songs but then the arrival of the Beatles and the British sound ruined this wonderful music and it changed to rock with its countless metallic, rap, and grunge variations. I doubt that there exist more than a dozen later epistolary, fire lighting, or deep rolling songs that I cherish.

I managed the equipment for my high school football team. There were 3,000 students at Curtis and there must have been 100 players on this superb team, which many years later went on to win the city championship. The coach, Andy Barberi, was so popular that the city named a ferryboat after him. He was a tough man but I did just fine. We traveled all over the place playing other teams. It was of very little real interest to me. I soon retired so that I could work and earn money for skiing.

I had some tropical fish. I repaired bicycles. I rode so much that I no longer enjoy pedaling. I used my hybrid English racer to commute to the city (from high on Staten Island); I stayed with Marvin on weekends. I took the now defunct electric ferry to Brooklyn and rode alongside the Belt Parkway all the way to Coney Island. I rode a lot. Cars did not come with seat belts and riders did not wear helmets. I never fell; I never hit my head. One day, when I was much older, I was carrying a bag of groceries in one hand and steering and braking with the other. I descended an extremely steep hill and could not stop. I crashed into a moving car. I was at fault but the driver was badly frightened; she asked me over and over again if I were okay. I was, and I apologized profusely. No harm was done to the car or to me. I stopped flying along much too quickly on two wheels and began to buy and repair automobiles. The moment in the summer of 1959 that I turned 18, the year that the state and city designate that one may legally drive (though in Iowa one may do so at 15, I think), I purchased a white, 1952 Ford convertible. Naturally, I drove around. Once, I picked up far too many people uptown and allowed some to sit up high on the trunk with the roof down. I could have driven through the streets, but, no, I chose to go to the Village via the East River Drive. The police were aghast; I received a $10 ticket for blocked rearview mirror and overloaded vehicle. I was really smarter than that! It could have been a disaster. Although this car was just lovely, I soon traded it in for the most wonderful vehicle I have ever owned, a yellow, 1954

Mercury convertible with a 1958 Thunderbird engine. It had power seats, power windows, and a hot-rodding power to help me win trophies had I cared about such things. I just drove it around town and back and forth to Vermont. On one return trip, I foolishly let Cliff drive. He was always part machine and he flew along public highways at a consistent 90 miles an hour. We were lucky to avoid jail.

The police were always there stopping us for imagined transgressions. Cliff had purchased a real Jeep (not one of the newfangled, luxury Cherokees); these were very rare in the City. As we drove up Pike Street, a cruiser pulled us over. Two officers (male and female) requested Cliff's papers. They looked at him; they looked at his license; they looked again and burst out in raucous laughter. Cliff must have been 20 but he looked about ten. I am sure they hurt his feelings and this resulted in many of the insane things that Cliff did, like getting his Jeep stuck on the Appalachian Trail! One day, four of us were tooling around the Village, and two officers lined us up facing a wire fence. One clearly stated that we should put our hands up on the wire and **not move**. How might one interpret this? Marvin decided to wander off. The officer yelled. Never defy someone who may be pointing a loaded weapon at your back. Police officers lead dangerous lives and every interaction offers potential harm. And some of these good folks are insane. Once, in the little bucolic Vermont town of Newfane, two state policemen (in a single cruiser) pulled me over because I tend to hug the right shoulder in order to avoid head-on collisions with all of those drunk, phoning, texting people heading my way. One officer came up to my window and asked for my papers. At that time, there was no unspoken convention that I had to stay in the car, so I got out. My officer did not say anything, but the man in the car shouted hysterically that I should get back in. He should have had a desk job. Or consider this trip to Ohio: I was enjoying the ride when at ten at night, the muffler on my little sporty Alfa fell off. I placed it on the front seat and continued noisily on my way. All too soon, an officer pulled me over. He did not like the sound of my jet engine. I explained that the mishap had just occurred and I could prove it. All he had to do was touch the muffler which was still extremely hot. As he bent over to do this, he unbuckled his weapon. Later that night, after amazingly locating an open garage that could make the repair, I asked him about this. He was being cautious in the dark with a stranger transporting a hot, rusty muffler. I had paid $1200 for this vehicle. I subsequently discovered that a pre-welded Abarth exhaust system from the manifold through the muffler and exhaust pipe, imported from Italy, cost $1000. I stuck with the straight barely muffled repair until sometime thereafter the Athens police forced me to modify it. This officer had an excuse: He observed that I passed an old age home every day, and I was perhaps annoying the sleeping residents!

My brother and I collected stamps. We had some United Nations materials, but eventually we specialized in mint US singles and plate blocks. It was a glorious collection which accreted slowly because of cost. As we moved back in time, the stamps became more and more expensive, and we had very little money. Don had no real source of income and I would have to steal some from my skiing budget, something that was very hard to do. Still it prospered. Some 25 years later, I sold it in Oklahoma City and split the proceeds with Don. This dealer got a good deal. (Well, of course. I recently sold a mid-1890s silver dollar to a dealer for $22. This particular coin, in mint condition, goes for $173,000!)

Marvin had built some hi-fi components and as my interest in classical music increased, I decided to do this too. I purchased a Wollensak reel-to-reel tape recorder, Eico amplifier and tuner kits, James B. Lansing speaker, large speaker cabinet kit, and semi-professional Rek-O-Kut turntable. I assembled the components and Marvin soldered the connections. I listened to classical music whenever I was at home, day and night, which annoyed my puritanical father, who usually preferred stoical silence. I bought a few records including the first two that Joan Baez made. My father must have seen them. One day, he was at a science conference and ran into a physicist bearing a Dr. Baez name tag. He mentioned that his son had recordings by a Joan Baez. The man replied, Yes, she's my daughter! (My dad was also at the AAAS meeting at which some sophisticated scientists poured blood or water on Edward O. Wilson's head because they did not like what he said in *Sociobiology*, despite the fact that he is correct!) I listened a lot, so much so that I tired of the normal orchestral repertoire and eventually turned my attention to chamber music, esoteric sonatas, duos, trios, and especially string quartets of which there are many by Haydn, Mozart, Beethoven, and Schubert as well as others whom I do not enjoy as much. Quintets and sextets, septets and octets are also among my favored works. The period I most enjoy begins about 1700 and runs through the later Romantics around the end of the nineteenth century. There are a few 20th century works I like very much, but serial composition (Webern) and moderns like Britten, Messiaen, Boulez, Crumb, Ligeti, Larson, and their brothers and sisters give me a headache. I do like de Falla, Mahler, and Poulenc, and even a bit of Stravinsky, Shostakovich, and Glass because they are merely recomposing the past.

I thought that I should play an instrument so when I was 14 or 15, I began private lessons. I chose the tenor saxophone, learned to read music, and practiced diligently. Despite the fact that this may well be the easiest instrument to master, I never got very good. In college, I tried the piano, but this too was a disaster. I learned the pieces, played them, and received a good grade, but I knew that what I was doing was inadequate. Later, bored with the saxophone, I bought a used clarinet and taught myself how to play. The sax covers two octaves and it is very easy to control the large reed; the clarinet manages four octaves, it is much harder to master the reed, and the fingering is complex. I gave up.

My single most consistent activity was basketball. From the time that I tossed a ball into the hoop during my eighth grade recesses until I stopped playing around 1995 because I feared getting into a fight with a student during very competitive faculty-student lunchtime games, I played a great deal: In New York City at schools and in tough playgrounds; at the University of Pittsburgh, where a serious fight erupted almost every time I participated; at the University of Oklahoma, where semi-professionals would stop by; in St Cloud; and even in Austria, where I managed to locate games. Lamentably, I never grew much: I am 5' 6" and rather skinny so one might imagine that chess would be a more suitable pursuit. I cannot rebound against someone who is six-six, but I have an admirable foul line shot (from 15 feet). When I see current professionals shooting foul shots at

50 or 60 percent, I am stunned. In the past, great shooters like Larry Bird or Oscar Robinson or Jerry West would get the ball in the basket with great frequency. Today, college players and even professionals often miss and the excuse that we now play better defense is just that, an unconvincing excuse for lack of skill (as the 2011 Butler/Connecticut NCAA championship game aptly demonstrates). I once shot ten in a row. I also have an excellent baseline jump shot (from about 18 feet. Additionally, even today, and bizarrely, I am very fast. I was an excellent defender, and stole the ball when unsuspecting players let down their guard. Shooting is like riding a bike: One never forgets. I can still run and play but no longer do, not because I fear the broken jaw and foot that I acquired in this non-contact sport but because I got out of the habit and in retirement have other things to do.

I bought books, lots of them. I haunted the many used bookshops that lined Fourth Avenue; I visited rare bookstores around the city. I was open to paperbacks, but I especially liked handsome special publications: The Limited Editions or Heritage Club volumes were fine and older things too. Large format overviews of Chinese art or facsimiles of well-known manuscripts caught my eye and somehow I did increase the size of my library. Books were quite inexpensive then but my income was proportionately smaller. As late as 1979, a first edition of Joyce's 1922 *Ulysses* could be had for $1,000. Recently one sold for 275,000 pounds (almost half a million dollars), which is perhaps the most absurd thing I have ever encountered. My wife collects too, but her only interest is subject matter. When I built our mountain retreat, I included two small libraries. And naturally we had many books at our South Burlington apartment. When Kira was five, she already had about 500 books. I would estimate that Terry and I own between 10 and 15,000 volumes. They are not mere decorations. We read them as well as library materials. We read a lot.

I had long been interested in the martial arts and taught myself some basic procedures, but without a teacher and frequent practice, I made no progress. When I was about 55, I would guess, I began to study karate (kempo) formally and was quite diligent in both the physical and mental aspects of the process. My senseis, a married American couple, both of whom were *san-*

dans (third degree black belts) in shotokan/ kempo, were extremely rigorous. All terminology had to be learned in Japanese, formal procedures were enforced as if we were at Marine boot camp, katas had to be perfectly executed, sparring was intense and dangerous, and testing included not only a physical routine but also an examination of historical, physical, and mythical aspects. The student had to know exactly where a specific one of the 750 odd pressure points is located, why one struck there, and what this means in terms of the air/earth/water/fire cycle. Breathing and recognition of the chi (force, energy) and other somewhat mystical elements made all of this more difficult than it would have been at Karate America, which one may find in dojos located in various malls!

I intuitively knew that my dojo was especially rigorous: Very few people would put up with the extreme precision demanded. Recently Kira, at ten, began to study taekwondo.

Before she started, I warned her that this was unlike any of her other pursuits, which are always performed under very liberal tutelage. So she knew what was expected, but then reality encroached: I am horrified by what went on at her dojang: High ranking youngsters banter, talk, chew gum, go through the motions, and are unsure of their moves. None of this ever occurred at my dojo, no matter how young a student was. There it took as many as ten years to move up the belt ranks. At Kira's dojang, one can reach black belt status in about three years, and age is irrelevant. In my studies, I only encountered two persons with black belts. In Kira's group, half the people, some of whom are still in grammar school, sport this normally extraordinary mark of distinction.

One day, I asked my sensei why a specific point on the top of the foot was a better place to strike then a comparable location off to the side; the sensei pressed down at this precise neurological juncture and sent an excruciatingly sharp pain through my body. When sparring, we chose partners arbitrarily. Every once in a while, I joined an older man (most of the students were in their twenties), whose solid commitment had resulted in a black belt (one that had real meaning); he sparred as if I were a menacing Samurai warrior, firing powerful hand and foot thrusts at me in rapid succession. Had one of these made contact, I would have been seriously hurt or even killed. I persevered and moved up the colorful belt ranks, but eventually gave up. I am not sure why.

This bothers me because it is one of only three things that I had set out to do but failed to complete. I did not earn a black belt (which in my dojo entailed much work with a bo [a big stick] and dual knives [had I wished to use weapons, I would have purchased a revolver]). I did not earn my second doctorate because I failed to write the required dissertation. And I have not completed the 50 state high points: I still have five difficult mountains to ascend; of the three tasks, this is the only one that I might still achieve (although now that I see what occurs in taekwondo, perhaps I will return to kicking and punching). Despite my age, I believe that I could successfully climb Rainier (WA), Granite (MT), Gannet (WY), and Moana Kea (HI); McKinley, also called Denali (AK), is another story. It is not impossible (I am still strong and motivated), but very dangerous, very difficult, and very trying. One must devote a minimum of three weeks to the climb, since cutting it short (for example, if one miraculously has perfect weather and days are not wasted waiting in a tent for the snow and wind to subside) will almost certainly result, at best, in a mild or severe case of acute mountain sickness; at worst, one may come down with pulmonary or cerebral edema, and death will follow unless one can descend very quickly. Perhaps I should just give up my quest.

Although I like animals (and therefore refuse to eat them), I have not had many pets. The fish I mentioned and occasionally during the summers a temporary cat or two, but these creatures never accompanied us back to the city. In 1966, just before I left for my second year at Ohio, I purchased a wonderful golden retriever puppy. The owner lived in South Londonderry (where I had attended the eighth grade) and happened to be the president of the Golden Retriever Club of America. I paid $200 for this purebred animal, which was a great deal of money for an impecunious student. Because I took classes and taught, I was gone for many hours each day. I sometimes returned to a free-running dog (He had pulled out his steel stake) or to a room filled with the remnants of Kierkegaard and Nietzsche (he had good philosophical taste). I finally had to give him away. When

Terry and I got together, we acquired a cat and then another and finally a third so that between 1968 and 1971, when we moved to Iceland, we had three producing animals. No one at that time proselytized against kittens. We had 53 of them and gave all but one away. In 2009, Kira decided that she wanted a kitten and so Silver is part of the family. We had hoped that she would produce some offspring, but that is now impossible. A few years later, we acquired another kitten. Tiger is adorable.

FIVE
READING AND MUSIC[1]

It would be foolish and undoubtedly incorrect to claim to have read more than many or most other human beings, but I am certain that I would give a host of contenders a good run for their money. I have been an avid reader since childhood. Toward the end of my senior year in college, I began to keep a record of my accomplishments. This catalogue, which covers almost 50 years, is devoted primarily to literary works. It contains some 2,000 listings. It does not cover school texts, technical treatises, and the work that I perused or skimmed while in libraries. It also does not separately list most of the thousands of items that I used to create my 600 published letters, reviews, articles, and (12) books. For example, the bibliography for *The Mountain Encyclopedia* contains ca. 250 entries; the references in *Authorial Ethics* number 300; and those in *Documentation*, 400.

I have read diversely and a great deal, as noted above: Anthropology (Benedict, Mead); archeology; art and architectural history (Gombrich, Panofsky); Biblical commentary (Rashi, but not in Hebrew nor Aramaic); biology; botany; chemistry (Pauling's *Nature of the Chemical Bond*); clinical trials; cliometrics; ethics (Spinoza, Bonhoeffer); ethology (Tinbergen); fiction ("The Shipwrecked Sailor," in Middle Kingdom Egyptian hieroglyphics);

---

1   I have a superb memory (though not as good as my eleven year old daughter's, which approaches the photographic). Except for counting the listings in my record book and the bibliographical entries in my three cited works, I did not consult any sources for the merely exemplifying list of readings and thus there may be one or two nominal spelling errors in evidence. A similar point may be made for my remarks on specific pieces of music, since I only checked three or four names; it is also not impossible that I am mistaken when noting a key signature, opus number, or Deutsch (Schubert) or Köchel (Mozart) listing, since all of this is recorded entirely from memory.

geology; hermeneutics; history; iconography; jurisprudence; karate (Funikoshi); law; memoir; mountaineering (Petrarch to House, Whymper to Wickwire); musicology (Stravinsky); mythology (Campbell, Eliade); nanotechnology; national literature (the works of countless authors representing countries in every part of the world, from the *The Odyssey* to *Der Zauberberg*—in some of the ten languages I have studied); neuroscience; optics; paleography (Mabillion); penology (Foucault); pharmacology; philology; philosophy (everyone); poetry (Dante, in Italian); psychology (Freud, Skinner); physics (Einstein—whose estate granted permission for me to read his technical papers, though I declined to do so); sagas; seismology; sociology; theory (from Derrida to Lacan, from deconstruction to semiotics); the Torah (in Hebrew); taxonomy; theology (Aquinas, Suzuki Roshi); urology; vulcanology (Thorarinson); xenotransplantation. I imagine that if I spent a few days on this I could create an enormous catalogue. In literature, I like and have read in all genres and their subdivisions: poetry, drama, serious and sometimes more popular fiction, mystery, science fiction, adventure, essays, criticism, theory. The only lacuna is romance.

As a teenager laboring at a law firm (discussed below), I recall the 23rd floor receptionist sitting alone at a desk at an infrequently visited entrance reading Dostoevsky's *Possessed* (also called *The Devils*). Although I was already engaged with serious literature (Melville's *Typee and Omoo, The Caine Mutiny* shortly thereafter), I was put off by this enormous novel. I still am. Nabokov is certainly incorrect in vilely denigrating Dostoevsky (along with Thomas Mann and Eliot), but the old Russian master is so verbose and goes off in so many misdirections, in *Crime and Punishment*, for example, that I cannot force my way through this, his most famous work. He is better in shorter pieces such as *The Underground Man*. I have the same trouble with Melville's masterpiece. I am working on my eccentricities, although I am going to live (and die) without any more Joyce and Proust: linguistic gimmickry and precious social interactions are anathema (*The Anathemata* is quite lovely though).

While I read I listened to music. At first, rock n roll and then classical compositions, day and night. In NewYork, WQXR and WNCN provided ongoing pleasure. From eleven PM until six the following morning, the latter station featured Bill Watson, the most knowledgeable, eccentric, and universally beloved host imaginable. One night, he played the entire Beethoven piano sonata cycle. When the last bars of the 32nd sonata concluded, he said, *Let's do that again!* And he did. He only played a handful of composers but those happened to be the ones I prefer: Bach, Handel, Haydn, Mozart, Beethoven, and a few others. No Telemann let alone Crumb; no Corelli, so certainly no Cage. I listened avidly, as I did in the Vermont mountains to WAMC, WMHT, and WFCR, each of which emanates from a different, distant city. Lamentably, every morning, WFCR featured the egotistical and annoyingly lethargic Robert J Lurtsema and for many years I suffered greatly with his inane comments. In Pittsburgh, it was WQED, which I liked so much that I volunteered at the station. Things were more difficult in unsophisticated locations such as central Oklahoma, but still I listened and cringed when the students from an Edmund college station bizarrely mispronounced Mozart or Ginastera. Later in Minnesota or back in southern Vermont, I switched to the music channels that come along with cable or satellite television, and I still prefer these, especially to Vermont Public Radio, which plays excerpts or very brief pieces, and follows contemporary programming theory which dictates that every composer, no matter how horrific, deserves airtime. Naturally, I disagree.

My favorite instruments are the harpsichord/piano and cello. I like all of the woodwinds and the flute as well as various percussion devices. But I consistently dislike the organ (except for the Bach *Toccata and Fugue* in D minor and the third Saint Saens symphony), and I cannot abide the entire panoply of brass instruments including the French horn even when it is articulating the four exceptional Mozart horn concerti. This surprises me too. Sometimes, as I read or write or edit, I get a very uncomfortable feeling. I look up and discover that some trumpet voluntary is polluting my environment and I turn it off. (Naturally, I may occasionally savor Brandenburgian horns or the Haydn brothers' two trumpet concerti.)

I guess that I listened too much because I got tired of the symphonic repertoire. After hearing Tchaikovsky's first piano concerto or Beethoven's fifth and ninth symphonies or his violin concerto two or three hundred times, I have them in my head. It is true that the orchestration is lacking but the essence is there for me to play mentally, like this: One day, I was returning home from Albuquerque, where I had briefly visited with my wife, who was working on a Master's at the University of New Mexico. I arrived at my seat and waited. Along came a man whom I had seen kissing an elegantly clad woman good-by in the airport terminal. He sat next to me and opened a score and began to study it. I was sorely tempted to sing the various parts in German, for I know them by heart (*"Zu Hilfe, Zu Hilfe, sonst bin ich verloren.... Der listigen Schlange...."* and so on) but my innate shyness and my horrible voice helped to control the compulsion. He was on his way to conduct *Die Zauberflöte*. I am certain that I mentioned that I was capable of torturing him, had I been so inclined; he undoubtedly would have changed seats! I know that I have this effect because my girlfriend R (whom you will meet toward the end of this memoir) had perfect pitch and would cover her ears when I sang because it was so painful to listen to a voice that is just "a little pitchy."

So, since I had progressed from the "Saber Dance," the 1812 and other overtures, and various popular capriccios, and tired of symphonic composition, I had to find something I liked more. Naturally, I could still enjoy a symphony and especially the four that Brahms had written later in his life because when I was much younger, I did not appreciate Brahms so I had not overdosed on his work. At about the same time, that is, some fifty years ago, a researcher discovered the score for Bizet's *Symphony in C*, which he had written when he was about 16, and which had long been lost. This is a lovely, lyrical piece of music that never fails to please. The same can be said for some of Dvorak's and Sibelius's symphonic works, Frank's *Symphony in D*, and Mahler's early, large scale compositions. Indeed, his *First Symphony* ("The Titan") contains a lugubrious adagio movement (the folksong "*Frere Jacques*") that under normal circumstances I would abhor, since I do not like slow movements, but which here is earth-shatteringly moving. In light of my previous remark and despite all of this contravening symphonic adulation, I turned my complete and rapt attention to chamber music. Mozart's early exquisite piano works are among my favorite sonatas: K. 289, 290, 291,

292, 293, and 294, the first six that he wrote, when he was still a child; 309, 310, and especially the ninth (311), my very favorite, are sublimely magnificent. The single greatest piano sonata is the second of Schubert's three posthumous pieces (D. 960), which contains four heavenly themes.

When I would attend a concert with R, she would announce upon hearing the first two or three notes, that the piece was in C-sharp or A-major; I had no idea what this really means, since I am tone deaf, but I knew the piece; she did not. This reminds me of the professional musician with whom I once talked. I was excited about a work she had played. I asked what it was and she replied that she had no idea: They put the score in front of me and I play, was how she put it. Other professional musicians have complimented me on my knowledge, some of which I have forgotten, but professional status is no guarantee that a person will have tenable opinions. A real genius (with 700 GRE scores), a professional violist, who played in quartets purely for pleasure, intensely disliked the entire Chopin corpus. This seems bizarrely impossible, since much of Chopin's work is extremely beautiful, especially his two piano concertos.

Duos and the wonderful body of work written for trios are most pleasing; I especially like two Brahms pieces (Op 34 and 40), but more than any other form, I like string quartets (and occasionally the piano variation). Haydn's amazingly replete collection, Mozart's "Dissonance," Schubert's "Tod und das Mädchen," and "Die Forelle," but especially Beethoven's 16 masterpieces. Numbers two, four, and six of the Opus 18, ironically, are stronger than one, three, and five (the odd numbered symphonies are said to be less lyrical and more powerful than the even numbered, and they are). I like the Razamovskies (Op 59, 1-3) best. But even some of the Late Quartets (Op 131, for example) as well as *Der Grosse Fuge* (which I am playing in my head as I write) offer the highest form of esthetic pleasure. The Schubert and Mendelssohn Octets are both exquisite.

The Bach cello suites are superb as are Schubert's magnificent Arpeggione sonata, now played on the cello, and the Dvorak concerto. Whenever a composer offers a broader and fuller role to the cello in quartets, I am grateful. Villa Lobos, who wrote a piece for nine cellos, carried his instrument into the Amazon basin, and Mark Salzman, who went to China in order to study martial arts and teach English, relates in *Iron and Silk* that he hauled his cello along when he went off into rural areas and played it for rapt listeners. The wonderful film, *Tout les Matin du Monde* presents an overview of the life of Marin Marais, whose cello exercises and compositions comprise the entire score; I wish that some Bach had been interjected (even though this might have been anachronistic).

There exist three magnificent works for clarinet, the instrument that opens Gershwin's *Rhapsody in Blue*, which as you might infer, is not among my favorite works, because I do not allow that America—or England for that matter—ever produced any world-class composers. So, must I apologize to Gershwin, McDowell, Copland, and Bernstein, not to mention William Schumann and Libby Larson or Purcell and Britten? No, I'll pass. I do, however, like Phillip Glass and Peter Warlock (a pseudonym for David Hesseltine). The first is Mozart's clarinet concerto (K 622), written not long before he died, while working on the Requiem (K 626), his last (uncompleted) piece. The clarinet soars liltingly and mellifluously transporting the listener (me) to another realm. As powerful are the two Weber concertos, perhaps the finest works ever written for this instrument.

The violin concerto is so important that someone drew up a list of the big four. I have never seen these ranked, but I would wager that most knowledgeable critics would place Beethoven at the apex and descend through Brahms and Tchaikovsky, and conclude with Mendelssohn. Two and three might be interchangeable, but the Mendelssohn would almost invariably come in last because it lacks the blustering grandeur that one finds in many of Beethoven's larger works, and the duration of the others. This is most unfair; in fact, the Mendelssohn concerto might just be the most extraordinary of this elite group. I was never very happy with its limited extent and exclusionary nature so I created a second group: Sibelius, Paganini, Paganini, Bruch. Apparently this too was inadequate because I went on to a third group of four, but only recall Winofsky, whose orthographically challenged name I can't spell.

I do not like the human voice and so I find a great deal of classical music and all of opera to be merely annoying. Nevertheless, there are perhaps 50 pieces that I like a lot: Handel's *Dettingham Te Deum*; some Bach cantatas, especially number 51 (*Jauchzet Gott in allen Landen*) and his *Magnificat*; the Vivaldi *Gloria*; *Il Barbieri's* famous aria; Villa Lobos's amazing sixth Bachianias Brasilieris, which is "sung" with sounds rather than words. In Pittsburgh, the moment I discovered that a fellow student came from Brazil, I mentioned the Villa Lobos piece. She must have liked me, because when she returned from her Christmas vacation, she brought me a version of this work that her mother had recorded.

The astute reader will have noticed a series of impossible lacunae: I have not mentioned any Spanish, Hungarian, Norwegian, Russian, or French composers. I have nothing against Arriaga, Liszt (excluding his bizarre transcriptions), Grieg, or Borodin and I am certain that I could dredge up some Rameau or Couperin that would be passably pleasing; but the two great modern Frenchmen, Debussy and Ravel, are not for me. The roiling sea, dallying faun, and dead Infanta are much too impressionistic: I do not like program music, unless Vivaldi is seasonally impressing or Tchaikovsky has been to Florence. I do well without any of *Le Six* (except Poulenc); I am, of course, grateful that they included a woman in this exclusive group, though few stations play Tailleferre and even fewer people ever heard of her. There are many others: Offenbach, Gounod (his *Faust* is exhilarating), Gliere, Glinka, Shostakovich (with his lovely repetitive Hebrew melody, and a son and grandson who lived next door to my brother), Kodaly, and Bartok and Nielsen (both of whom I do not like), Tartini trilling with the devil, Rossini, and so on, almost interminably.

Different people have varying skills. I once asked two "musical" colleagues to take "Hauptman's Epigraph Quiz," which consists of 25 catchphrases that have been attached to compositions; they range from easy (Eroica) to difficult (Antarctica). Fred B, an excellent pianist who can recall almost any musical theme, managed to identify three of them, whereas Mark, the author of a Verdi bibliography, got 23 of the 25 correct. My skills lie somewhere in between: I can hum only a hundred or so themes or often indicate the next movement when listening to a piece, and I probably would have been able to identify only half of the entries on someone else's quiz.

SIX
SENSITIVITY

I am extremely sensitive, not to offenses or insults, breaches in etiquette or the continual stupidities of humankind, but rather to the difficulties and challenges and miseries that the world and its inhabitants inflict. I have long humorously described myself to people I meet as a woman in a man's body, under the patently false assumption that woman are more sensitive than their male counterparts, a general cultural myth assumed perhaps because females enjoy romantic comedies or cry more easily; but men have been *acculturated* to be tough and stoical, to emulate army commandoes or navy SEALs, despite the beer they guzzle, the cigars they smoke, and the fat they carry around. Screaming hysterically at a Giants football game does not mean the male is inherently less sensitive or the female more so. At any rate, I am compassionate and empathetic. I care much too much in a world controlled by rapacious predators, demagogues, and people who enjoy harming and killing, who attend bull and cockfights or who produce violent songs, books, and films that desensitize young and old alike. Chainsaw massacres, hunger games, or gangsta rap, works with few social, cultural, or literary redeeming qualities, inundate the American landscape and since the US is the most powerful country in the world, in virtually every category, we export this dross, and I imagine that young Indonesians or Mongolians walk around with their oversize baseball caps askew chanting about their bitches and how they hate their mothers.

    I recall that, as a youngster, I found a mouse in a trap that my dad had set. I was saddened to the point of tears. And it was not long after, that I became an ethical vegetarian which solved a big problem: I no longer had to feel guilty about killing, eating, or abusing animals. Natural and human harm continue to haunt me. Tornadoes, cyclones, and earthquakes and hunting, fishing, crime, wars, and genocides take an unimaginable toll. I am especially affected when children are involved. Even the sociopathic author Louis-Ferdinand Céline maintained that despite human depravity, we can still place our hope in children. I brood for weeks, months, or decades about a particularly egregious act, and there are many more than those that come immediately to mind.

It is perhaps because I am so sensitive that I overreact. I worry, I fear, I enjoy. I inherited my worrying nature from my mother who continued to fret about her children until she was almost 90. And I always know that bad things are coming so I fear: insane drivers, rapacious muggers, and in the mountains, grizzlies, cougars, lightning, and avalanches (in that ascending order)—and all with good reason: Those who do not fear real danger in an alien environment perish very quickly. This is not unwarranted or neurotic anxiety, but rather a healthy fear of things that harm and kill the unwary at an astonishing pace. Every year car accidents have killed as many as 50,000 Americans, 5,000 of whom die because of long-haul truck drivers, and avalanches harm or kill hundreds of skiers and mountaineers sometimes in groups as large as twelve. On the other hand, I can easily reach a state of rapture listening pointedly to Beethoven or Schubert or Brahms chamber music; indeed even the great rock n roll classics of the '50s capture my unwavering attention. And I can have a very powerful reaction to extremely demanding physical work. Anyone who does something, whether digging a ditch or setting pipe on an oil rig, will complete the task and feel a sense of accomplishment. Indeed, the work can become its own reward. But there is more. When I hay (pick up heavy bales manually and transport them to a barn often in oppressively hot weather), haul maple sap though the snowy woods, or frame or shingle a house, I occasionally experience a sense of pure ecstasy, something akin to the runner's high (which I have never managed, despite the tens of thousands of miles I have run). This feeling is far more powerful than that which I usually garner from my big mountaineering adventures. When I try to impart this (not the experience but rather the fact that I have had it) to my brother, he counters ideologically with the news that humans have spent their existence remediating labor, creating tools and devices to rid of us hard physical work. But since he has never done any and since he has never had the experience that I am describing, his reaction is akin to a celibate priest lecturing on coitus or a man, even one who has participated in a couvade, claiming to understand the physicality of pregnancy and childbirth. He is in the dark and I am floating-along with the low hanging, foggy clouds that shroud the snowy mountains early in the morning as I ascend over and over again. Very infrequently, I pass above them and come out into the startling sunlight with millions of snow crystals glistening in the air. I stop and encapsulate them. Others speed away on foot or skis or snowboards. They miss the real joy.

SEVEN
HEALTH

One afternoon, as I sat on the living room floor eating a hamburger, I had a revelation. No precipitating occurrence preceded this. I had not murdered a dolphin, enjoyed a bullfight, run over a turtle, read *The Jungle*, or visited a slaughterhouse. (I had of course been to butcher shops and Marvin's father's wholesale provisioning company and I had seen chickens killed, but these things played no role in my decision.) I had never given any thought to an alteration in diet. The idea simply appeared and I liked it. I became a fanatically strict vegetarian and regardless of situation have never really varied an iota in thought or deed. Because my parents maintained a strictly kosher home, I have never eaten any seafood. And for almost 60 years, I have not eaten any meat, fish, or poultry. I do not consume food that contains animal products, e.g., jello (gelatin) or Worcestershire sauce (anchovies). At first, all I cared about was the ethical component, which mandates that humans should not harm or kill for pleasure. (And so I also try to avoid animal skins, hides, taxidermy adornments, glues, medicines, ad infinitum). But as the years went by, I additionally came to care about my health and so, for the most part, I avoid eggs, butter, cheese, and other arterial cloggers. I do drink (skim) milk and consume its progeny (yogurt and ice cream).

At home, this is easy; in restaurants, it is cumbersome; in people's houses, it is embarrassing; and in other countries, it can be impossible. Since I have traveled extensively in most parts of the world, I sometimes went hungry. After visiting some 30 countries, I knew what to expect in Japan, where fish insinuates itself into everything, but it was still very hard. For 10 days, Terry and I ate Seven-Eleven bread and cheese. Yes, I know, restaurants have little dumplings and rice, but I failed to locate uncontaminated versions of these items.

I have always been extremely hyperactive. Even today, at 71, when I should be rolling around in a wheelchair, I still run extremely quickly (a five minute plus mile a few years ago, which was a big mistake because I could not catch my breath), jump off things, climb big mountains, and work hard (12 straight, uninterrupted hours a few summers ago, repairing a wall and rebuilding a deck with heavy pressure-treated lumber). As a youngster, I played basketball and skied so I was always in fairly good shape. When I was 18, I decided to work out at least once a week. As I aged, I incrementally increased this to twice a week, then three times. I now run out of doors, year round, regardless of temperature, rain, or snow, seven days a week. (I refuse to run in lightning storms and when I am ill, I stop, but this occurs only infrequently.)

I have been lucky in love and genes: My parents were very attractive and fit. Before I was born, they played tennis, and my dad did daily exercises at home. Lamentably, I never liked the way I looked, but I appeared amazingly young. When I was in my early forties, a fellow ball player stared at my naked body in the University of Oklahoma gym locker room. How old do you think I am? I wondered. He replied, twenty.... I cut him off. (At 65, my father looked 20 years younger.) So, I am fit and look okay. Exercising, eating carefully, never using drugs (even when I lived in Greenwich Village during the sixties), never smoking (except for a few times as a youngster), and never drinking beer, other alcoholic beverages, or coffee probably have something to do with this.

Figure 6: Hauptman's Parents around 1960, and in 1986

Just a few years ago, when I was 66, Fred (who is 20 years my junior) and I once again climbed Mts Baker and Adams, both located in the Pacific Northwest. These are admittedly difficult and dangerous climbs, but physical strength and skill are less important than desire and motivation. Accompanying us was Fred's girlfriend's son, Mike, who at almost six feet and 180 pounds and with a body honed through cross-country racing should have flown right up the glaciers or snow fields. On Baker, he simply could not continue despite Fred's incessant prodding. On Adams, he got a very bad altitude headache and could not take additional medication. We turned back on both mountains, a most unpleasantly ironic twist, because I was as strong as a Nepalese yak on that trip. Some subsequent climbs of these two peaks were real disasters.

When my father decided that his two year old son had to replace his toys or lose them, he created a conflict that resulted in a stress-induced ailment for which I obviously had a genetic proclivity and that has haunted me, on and off, all of my life. Sometimes, I am fine, especially when living alone in some Florentine pension or Appalachian apartment. But at other points, I have been so ill that when I entered the doctor's office, he began the conversation by saying, "You belong in a hospital!" I started using topical medications when very young and continue to apply these often useless (though palliating) unguents when necessary. When I was about six or seven (again), I spent some time in a hospital. I was given injections of ACTH, which stimulated my pituitary gland to produce cortisone. It worked. But I could not spend my life institutionalized. By 1967, when I had some bad luck with a female, I was able to ingest cortisone, which is very effective temporarily or permanently depending on what is happening psychosomatically. I dislike medicines and try to avoid them, but over the years, I have used Prednisone many times. For the past six years I have also taken one or two daily vitamin pills.

EIGHT
WORK

I started working when I was a very young child. My father believed that hard work and responsibilities built character. (Don't forget that John Kennedy—foolishly—worked while at Harvard!) During the course of my long life I have held 32 diverse jobs: I stacked (groceries), bought (monographs), delivered (money), shelved (books), unpacked and wound (watches), sorted (foodstuffs), constructed (retaining walls), served (food), maintained (bungalows), taught (many subjects), palliated (welfare recipients), hayed, sugared, sanded (hulls), washed (dishes), pruned (orange trees), edited (journals), wrote (and published more than 600 items). I began my first formal job (with working papers) at the St George branch of the New York Public Library during my sophomore year of high school. I shelved, and retrieved books from a holding area. Naturally, I worked after school for a few hours every weekday; the pay was 95 cents an hour, but it added up: I managed to buy a pair of Head skis, Cubco bindings, poles, and boots, as well as an orange sweater, which I still own. I also went skiing. I did this again when I returned for my junior year. I like books so I enjoyed the job.

Before he left to teach in college, my father had begun a coop program at Curtis High School, and as a senior I decided to participate. One week, I went to school and doubled up on English, history, and law. The next week, I worked full-time in the City at Angulo, Cooney, Marsh, and Ouchterloney, a large firm that specialized in testate law. (A friend held a similar position nearby at Dewey, Ballantine, and Palmer, the senior partner of which was the former New York State governor and the ostensible winner of the Truman/Dewey presidential election, at least according to the premature headline. I once saw him pull up in his big limo. The firm recently imploded.) The associates were normal people who probably worked hard for their middle-class remuneration, but the partners were extremely wealthy. They were secreted in their offices so that, for example, during the entire year, I encountered Angulo just once, when I entered his enormous empty office in order to have him sign some papers. My job was to deliver money, checks, and documents around the city. It was fun. I made $33 a week, and went skiing.

At about the same time, I tried my hand at healing at a small, rural New York City hospital. There were still dairy cattle on Staten Island so the hospital was no surprise despite Manhattan's enormous Mt Sinai, St Vincent's, and Bellevue. I was in charge of ambulance care. The other person did only two things: drive and carry one end of the stretcher. When not rushing around picking up the ailing (I never got to do this), I spent my time in the emergency room acting as an orderly. There was no preliminary training at all. This is incomprehensible, since emergency medical technicians are now so skillful and their portable equipment so sophisticated, they can probably perform brain surgery. The first patient to make it to the ER during my watch was a hoodlum who did not do well with his minor injury. Next came a howling child and a distraught mother. The little boy was not badly hurt but he made a great deal of noise. The West Indian orderly said, *Come on over here, mon, and I'll show you how to shave his head.* I took one look at the tiny cut (which did require stitches) and knew that I would faint so I ran out and sat on the curb attempting to recover. I then reported to the head nurse with the sad news that I could not work in the medical field. She caringly replied, I knew you couldn't!

After high school graduation, I spent the summer at our Vermont house and then immediately located a job at Helbros Watch Company. I wish that I had looked a little harder because this was neither challenging nor enjoyable. I used the weekly paycheck of some $50 to ski. At the end of the year, I left to go to college, where I dedicated my time to studies, but every once in a while, I held a part-time job: I cut out articles at a newspaper clipping service and I worked at the post office at Christmas. During two summers, I did all of the maintenance work at a Hyde Park bungalow colony. My responsibilities included mowing, plumbing repairs, shuttling children, tidying, and restructuring cesspools (which, lamentably, are very different from septic tanks). During my final semester, I worked full-time in the post office. This was a wonderful job. I unloaded trailers filled with heavy bags of mail and divided parcels by flinging them into distant laundry tubs. I boxed up mail. Once a week, I delivered packages (to the Empire State Building). One day, a young man playfully threw a one pound sample bag of Gold Medal flour into the air; it sailed down through his outstretched hands and burst open. He scooped the pristine flour along with the rubbish and filth found on the floor back into the bag and sent it on its way. Although one is constantly observed in large US post offices (including in bathrooms), he was not reprimanded. I retired in September and returned to Europe.

After I got back from Italy in the late fall of 1967, I happened to be wandering northwards in lower Manhattan; I looked up at a building and noticed that it was the headquarters of the city's social services agency. I walked in and asked a security guard when the next civil service test for welfare investigator was to be offered. He replied, Right now, and handed me a pencil. I guess that I passed the test because a few weeks later, I was working for the City. Hundreds of potential artists, writers, and opera singers were housed at the Greenpoint center in Brooklyn. Our appointed task was to visit welfare recipients (the poor, the retired, the overly burdened with offspring, the addicted), check rent and utility receipts, and offer some solace. Upon returning from the field, we orally recorded the data from our visits; this was then textually documented by a professional typist. Some of the files were enormous, since these folks had been receiving supplemental payments for decades. A few people just abhorred work. I did go into the field (while some of my colleagues went home), and I did speak at length with mostly good but underfunded recipients. I did not bother with receipts. I cared more about the emotional distress of, for example, the young woman who was impregnated (with her consent) by a prisoner at the facility at which she worked, and therefore lost her job. She wanted to have the baby at home but her middle-class parents insisted that she pay rent, so the welfare system came to her aid. I spoke with this depressed young woman for hours. Most of my interactions with clients were similar.

There must have been 200 caseworkers at Greenpoint, but only one medical overseer whose job was to evaluate special needs for the medically challenged. She was very begrudging and obstinate. I had an elderly, blind client whose daughter-in-law had thrown her out and who was rescued from the sidewalk. I requested and argued for a phone. It was like pulling shark's teeth, but I succeeded. Much worse was the advice given to the mother of a child with Cystic Fibrosis: Instead of paying the woman's enormous electric bill (which powered a life-saving device), she was told to use her ongoing clothing, furniture, and appliance allowances to cover these costs. What was she to do when she needed a new refrigerator? After many months, I received my official case load, which consisted of derelicts living in Coney Island boarding houses. I knew that this would finish me off and I would end up hospitalized. I quit and returned to Ohio. This entire process is unnecessary, almost a sham. It is as if those people who incur Medicare payments were visited monthly by case workers. What a bureaucratic waste of money!

I worked at Ohio University as a teaching assistant, a job I held from 1965 until I completed my doctorate in 1971 (with a single year off for study and work elsewhere). I taught freshman composition and sophomore literature classes. I began with full tuition remission and $2200 a year and worked my way up to $4500 (which included an extra class assignment). In any case, it was always enough to get by, even after Terry and I married.

There were many other minor jobs. After earning my doctorate, I spent 10 months at a hardware store, and then managed to stretch my savings and unemployment insurance benefits so that they supported us for four more years. One summer, I worked for an Ohio dairy farmer. We put up 10,000 bales. When stacking the hay high up just under the barn roof rafters, the temperature must have been 120 degrees. It was a pleasure. I had begun my haying career in about 1950, when my dad and I would help our neighbor put in his crop. At that time, we worked with loose hay, doing some raking by hand, then loading it on a truck with pitchforks. The barn was equipped with a hoist that could lift a large amount of hay off the truck and deposit it in the loft. Eventually the Newells bought a baler and we switched to these smaller units. I continued to help the Newells almost every summer and do so even today as a neighborly gesture. I like hard physical work, and that is my only reward. This is also why I helped the same people make maple syrup (in the traditional way), whenever I was in southern Vermont during the winter. But for this, I did get paid. There are many tasks involved in this process, including the tapping of trees and setting up buckets or attaching lines to the spigots, but the hardest is undoubtedly gathering the sap and carrying it to a portable tank. Walking through the woods on snowshoes carrying 90 pounds of very cold, sloshing liquid is extremely tiring. The sap then flows gravitationally into a large holding tank located at the sugarhouse, and from there into the enormous pan set on top of a furnace, which must be carefully stoked on an ongoing basis with four foot logs. At the correct viscosity, the syrup must be removed manually, filtered, and, while still hot, placed in containers ranging in size from a half pint to a 30 gallon drum. All of this must be done at very precise times: When the March or April temperature drops below freezing at night and is followed by a sunny, hot day, the sap flows out of the sugar maples almost as fast as water pours from a pressurized fire hose. Thus, one must work long, hard hours to avoid wasting this valuable commodity, 30 gallons of which will yield one gallon of maple syrup in the early part of the season. As time goes by, the sugar content decreases and by mid-April, one requires 60 gallons of sap to produce a gallon of real syrup. This is why it is expensive, and the result is that IHOP and other restaurants now offer diners synthetic versions unless Vermonters are willing to pay for the real thing. I imagine that Texans neither know the difference nor care.

Figure 7: A Young Bob Sugaring, by Terry Hauptman
Figure 8: Kira and Bob, by Terry Hauptman

In July 1980, I accepted an appointment as the Humanities Librarian at the University of Oklahoma. I was welcomed by 30 straight days of 105 degree temperature. I lasted four years. I left a job, colleagues, and an institution I liked very much for the much cooler environment one finds in central Minnesota, where one day I walked to work at 40 below zero, real temperature; I only felt the cold inside my eyes. This weather depression broke the Minnesota record: 60 below up north and 80 below wind-chill factor. The police announced that anyone caught driving a vehicle in two northwestern counties would be arrested! I stayed at St Cloud State University (SCSU), where I worked in the library and taught undergraduate and graduate classes in the humanities and social sciences, for 21 years. I retired in 2005.

When I arrived at SCSU, I was dismayed to discover that I had to join a union. I not only do not dislike unions, I think that they are the reason why America has prospered, but I do not like being forced to do anything. So I became a fair share member, which meant I could not serve on university committees. Eventually, I realized the stupidity of this position and took up full membership. I judged various competitions, awarded monies to faculty members and students (one of whom went on to win at the national level), and chaired both the graduate and grievance committees. I also do not like unnecessary meetings and bureaucracy, and so I tried to keep things lively and interesting. The grievance work, which entailed dealing with provosts, presidents, lawyers, and many colleagues who had been sorely harmed by both administrators and faculty peers, was most disheartening and stressful. Nevertheless, I persevered for many years, and helped save many people. Lamentably, I also failed in a few impossible cases.

Throughout my academic career, I have written and published reviews, articles, and books, and edited collections and journals. Many academics write a bit but for me this has always been a second simultaneous career and I have been very productive: My bibliography of some 600 items runs about 25 single-spaced pages. I continue to write and publish and to edit the *Journal of Information Ethics*. I was also frequently invited to speak at conferences and workshops, and even today, I occasionally present a lecture at a university or elementary school!

NINE
INTELLECTION

If I had been in Cambodia or Vietnam or China at just the wrong time, I would have been shipped off to a reeducation camp (if lucky) because, although I can do almost any type of work manually or with mechanical advantage (driving, hauling, milking, plumbing, disking, planting, weeding, building, deconstructing, repairing mechanical and, to a limited extent, electronic devices, photographing, lithographing, and so on), I consider myself an intellectual. Since I was 16 and asked my father to accompany me to an independent New York bookstore in order to buy a large number of classics (*Crime and Punishment*, *Moby Dick*, *Andersonville*), I have spent my time reading and studying. My goal was simple: I wanted to know at least a little about everything, even subjects that I find boring like baseball, economics, or chemistry. I never cared about in-depth knowledge in a very specific subject such as the retina, Louis-Ferdinand Céline, or chess. I am a dilettante and happy with my achievement. I do not like trivia at all, but I am able to answer a high percentage of questions when monitoring quiz shows such as *GE College Bowl*, *Who Wants to Be a Millionaire*, *Jeopardy*, or *Cash Cab*. And I **am** smarter than a fifth grader! In fact, in 1964, I had an audition for *Jeopardy* and did very well, but because at 23 I looked as if I were ten, I was not chosen to play.

I have tried to study and learn about everything: the broad sweep of the humanities in some depth, the social sciences, and all five of the hard sciences. Naturally, knowledge accretes and what I might have learned of basic biology almost 60 years ago is now inadequate. But I have subsequently concerned myself with a bit of molecular biology, ethology, natural history, and taxonomy, so I have some notion about how things proceed. I cannot sequence Komodo dragon DNA, but I really do not want to. I am much too busy learning about cosmology, I Ching nonsense, identifying wild flowers, or the difference between sculpting in clay, carving in beech, or working in marble. I like to memorize things: the Mohs scale (in part), the Popes (much too long and repetitive), the US presidents (I am not sure why I gave up), Mozart's Köchel listings (my brother claimed that I had memorized all 626 of them, but that is untrue), instructional classwork, lectures, and speeches, all of the Nobel laureates in literature and the dates, vocabulary (in ten languages), phrases and the numbers one through ten (in perhaps 20), poetry (though I cannot compete with Emily Dickinson who knew some 4,000 lines in Greek and Latin), and the *Inferno* in Italian (progress here is slow and I may not finish this project unless I live to be at least 500).

One of the ways in which I have learned many practical applications is to stop to carefully watch people at work. When possible I would engage the lineman, farmer, cantaloupe irrigator (who spoke only Spanglish), machine operator, or store clerk in conversation and then when pressed with a real necessity ask a precise question (about phone line installation in the wild, ground rods, 220 volt wiring, or conduit burial). There is a certain wisdom to the old joke, I love work; I can sit and watch it for hours. I know a great deal because I have observed workers in both a public environment as well as in some highly secretive locations such as the Federal Reserve Bank vault in New York City. (I am amazed that they still allow visitors here, even though this private space is crawling with guards carrying submachine guns, which is truly bizarre: Whom do they plan to shoot?)

I have been an inveterate and diverse reader all of my life. But I have gone through three extended periods in which I specialized. In high school, I read all of the available accounts of escape from German prisoner of war camps, and then extended outward to related topics. Around 1970, I read an extraordinary amount of material in comparative literature, theory, and criticism for my doctoral exams. And from 1990 onwards, I have concentrated on the extensive mountaineering literature as well as information ethics. I have always read other materials as well both at home and in frequent visits to public and especially academic libraries all over the world—from Amherst and Williams

Colleges and the University of Minnesota health sciences branch to the Akureyri public, not far from the Arctic Circle. I visit these collections with some frequency in order to read periodicals (journals and popular magazines), even today when they are often available to me via the Internet.

I studied; I learned; I began to write. Naturally, I had written papers and exams all through school, but I wanted to publish my work and I recall the real joy my second journal acceptance produced. I had already published a volume of poetry, an article, and some book reviews, but this was serious literary commentary. We were in Pittsburgh, and I submitted a short piece to *The Minnesota Review*, which at the time had a Marxist slant with which I do not sympathize. At any rate, I howled with pleasure when I received the acceptance. I continued and have managed to produce perhaps 650 individual items, some 600 of which have appeared in legitimate, mainstream magazines, journals, and books. Somehow I have wormed my way into the letters columns of *Scientific American*, the *New England Journal of Medicine*, different sections of *The New York Times*, *The Chronicle of Higher Education* (perhaps 25 times), and the primary or review sections of countless other periodicals.

I have written and published 12 books. Many deal with technical subjects such as ethics and librarianship or technological applications concerning the storage and retrieval of information, but some are aimed at a popular audience and are extremely enticing: *The Mountain Encyclopedia*, with its 500 images, 400 in color, or *Grasping for Heaven: Interviews with North American Mountaineers*. Some of these books were collaborative efforts and I have worked well with my coauthors most of whom have contributed their fair share. Naturally, there are times when one person may do more real work than another, but things tend to balance out. I have never had an altercation with my countless coauthors (many of whom were colleagues whom I invited to work with me in order to help and mentor them). That is because I was invariably the senior author and the person with the idea, but I always placed the other person's name first. This way, there could never be any animosity generated. Infrequently, the coauthor refused this honor. I never offered to "coauthor" a piece with a graduate student who had produced an excellent paper. I may have suggested that he or she try to publish it, but I did not want my name associated with something with which I had little to do. (This is a time-honored means that academics use to increase their productivity; it occasionally may be warranted but generally I frown upon it.)

I have published in four general disciplinary areas: literary criticism, library science, ethics, and general interest. Although my favorite works fall under the fourth rubric, my real contributions have been in the area of information ethics, a term that I coined, and which has slowly grown (and altered conceptually) to encompass all disciplines. In 1992, I founded the *Journal of Information Ethics*, which I continue to edit. As far as I know, it is the only publication devoted exclusively to this specific (if broad) subject.

TEN
TRAVEL

With few exceptions, the only traveling I ever did, prior to going to Europe, was between New York and Vermont. My father took a direct route, slowly wending our way back and forth between homes. There was no stopping for tourist attractions, eating, purchasing, or motels. We always wanted to get where we were going as quickly as possible, although my father drove very carefully, as befits a meticulous person. When the police stopped him, it was to give him a ticket for driving far below the speed limit. And my mom's early domain was also extremely proscribed. As far as I know, she never left New York City except to go to Holyoke once or twice, where she worked in a summer camp and to take a bus to Washington, when the DAR refused to allow Marion Anderson to sing at Constitution Hall. (She purportedly also visited Tennessee.) My parents were New York liberals though not at all sympathetic to Communism or even socialism which was so typical of their peers. Once I boarded the Queen Frederica, I was unable to stop traveling. I have been in every state, some, hundreds of times; most Canadian provinces (including Labrador and the Yukon); and ca. 40 countries on four continents. I have also been very close to Central America, and a long stone's throw from both North and East Africa: close enough, I think. I only lack Antarctica, and I should go soon. I am guessing that with its cold, snow, penguins, and paucity of humans, it may turn out to be my favorite continent.

In December of 1962, Vinnie, Garret, and I drove in Vinnie's brand new, $900 Volkswagen through Nuremberg and East Germany to Berlin. We visited East Berlin, entering at Checkpoint Charlie, and were dismayed to see that the Communists were not as successful as the propaganda had led us to believe. It was a mess, and Humboldt University was closed because there was no coal to heat the buildings. The border guards confiscated much of the material we had picked up at the new Museum of the Wall and the considerate Volkspolizei stopped us on the Autobahn and stuck their submachine guns in our faces. We escaped to Hamburg, where I purposely annoyed my good friends, and so they threw me out of the car. I went to see *Taras Bulba;* they went to hear the Beatles, long before Ringo was the drummer. It's okay: I never liked this annoying, self-indulgent quartet, which enjoyed repeating "Hey Jude" ad nauseam.

Otherwise, I skied, unlike my fellow classmates who did a great deal of traveling (to Spain or Sweden) during the school year. On the way home in June, we visited Paris and London and embarked in Southampton on a lovely ship that made the Atlantic crossing, through a tumultuous storm, in just four days. (You will recall that the Queen Frederica dallied for 14.)

Then came Innsbruck. I bided my time until the break in April, when I bought a highly discounted train ticket for about $30 for Basra via Baghdad. I took a tiny shoulder bag filled with necessities and set out for a two month trip that covered Yugoslavia, Bulgaria, Greece, Turkey, Israel, Austria, Czechoslovakia, and Germany. I learned many things (men with guns sleep in youth hostels; young people are generous; hitchhiking is impossible in Greece; eastern style toilets are unsanitary; train travel in distant Turkey is worse than imprisonment; people mistreat their relatives; and it is possible to save lives by being smart). I met Doug at the American Dershanazi in Istanbul. One day, a Pakistani fellow was declaiming to a large group of people that Hitler was right. Doug jumped up but I grabbed him and explained that, despite his father's advice that if the Jews had stood up to Hitler, there would not have been a Holocaust and despite the obvious fact that Doug was a strong wrestler, he would spend the next few years in a Turkish prison, if he manhandled this apparently popular orator, whose audience consisted exclusively of young men from various countries. He calmed down. A few days later, we were wandering around near the water at five AM. We wanted to see the drawbridges rise. We had the bad luck to encounter an aggressive Turkish drunk; again I suggested that physical retaliation is a bad idea in a land where the *Midnight Express* would, at some future time, scare viewers out of their minds. I saved him again. (Having grown up undernourished in a tough area of New York, I knew how to avoid dangerous confrontations; this knowledge has served me well.)

The Dershanazi (a youth hostel) was located directly across the street from Hagia Sofia, at one time the church with the world's largest dome and now a mosque. I stayed in Istanbul for a long time, so I got to see this architectural wonder frequently. And nearby is Topkapi, formerly, the royal harem and now a wonderful museum. It has been almost 50 years, since I wandered through its rooms but I recall the layout perfectly. And clearly visible nearby is the famous Blue Mosque. For some truly inexplicable reason, I never walked over to it.

But I did take myself down to the street on which Istanbul's brothels were located. (This was the same type of sociological research I performed when wandering around the enormous and confusing enclosed bazaar and exiting far from my entry point.) The street had two swinging saloon doors barring the way, so that one had to enter as if he were heading into a bar, with Billy the Kid, for a shot of whisky. The houses were messy and sordid, and, as I understood matters, Turkish men were not allowed to avail themselves of the provided services. Since there were comparatively few tourists in Istanbul at the time, I do not understand how these folks were able to earn a living. Innsbruck, located in a country that disallowed divorce, had two brothels; Amsterdam's offered views of the lovely ladies, who sat behind large glass windows that passers-by could hardly ignore; and Hamburg is home to the infamous Reeperbahn. One wonders in vain how and why these many conservative countries offer government sanctioned prostitution, whereas the United States with its constitutionally guaranteed freedoms, has allowed a handful of puritanical blue bloods to at least attempt to control sexual mores and activity. Naturally, they fail, despite the large official sign I once saw on a San Francisco Street: Johns Lose Their Cars. Why do we put up with such evil? Naturally, being of a rather fastidious nature and having embraced feminism long before the First Wave rolled through America's conscience, I have never indulged in contractual sexual activity.

Doug and I took a ship from Istanbul to Haifa (the captain refused to stop at Nicosia, Cyprus because of the fighting), and hitchhiked all the way to Eilat. The hostel was full, so I slept on the beach with the Red Sea lapping at my feet, and the lights of Aqaba, Jordan in the distance. Doug was unencumbered but I had to return to school, so I left. I caught a ride in a truck and after about 30 miles, the driver said shalom. This was a very big shock because his shalom came in the middle of the Negev desert far from Eilat and about 100 miles from Beersheba, the next real city. He was making a delivery to King Solomon's copper mines and I was now trekking alone in the inhospitable desert. I trudged along. Suddenly, a vehicle hove into view, something that happened very infrequently. The driver stopped his jeep and I opened the door. Instead of welcoming me, he yelled, What are you doing out here? Don't you know that Jordanian horsemen come down out of those hills and shoot people? I just turned in my guns! Jossi was a Belgian studying in Israel. He gave me a ride all the way to Tel Aviv, lent me his Americanized Israeli girlfriend, found another, and we spent the evening in a smoky and noisy club. I hated it. I headed northwest.

I had excellent experiences in Athens, Thessaloniki, Vienna, and Prague, where I discovered that Kafka was not exaggerating. I almost always get my visas at borders, but for some reason, when in Vienna, I went to the Czech embassy and obtained a visa. It was good for three days. When my time ran out, I wanted to stay on, so I went to the appropriate agency to renew. I banged on a locked door, but it was wrapped in thick foam and covered with a leather-like substance. No one could hear me. Then someone more knowledgeable in the ways of Communist bureaucracy knocked on the window and rushed up to the interior door. I followed him in. I did not bother to mention that for some time we have had door bells in the free world. Prague was perhaps the most westernized of the Communist capitals, and it was there that I was able to obtain strawberry milkshakes. While waiting in line at the embassy in Vienna, I asked someone if he were driving to Prague. He said yes and offered me a ride. In the little vehicle was a young, perseverant American woman. She had picked up a puzzle ring, which consists of three or four interlocking, oddly shaped bands. Once these are separated, it is extremely difficult to reassemble it, if one does not know how. She spent the entire trip from Vienna to Prague working on it and finally figured it out. I must have been impressed because I got her name and address, which I kept for some 25 years, but sadly, I never saw her again. I returned to Innsbruck and my studies.

The second short semester proceeded and one day, I was sitting at my desk writing to my parents. I knew that my 17 year old brother was traveling in Europe but he was far from Austria. I was in the act of writing that he probably would not show up for two or three days, when there was a knock at the door. I was wrong. He stayed at a fancy hotel using his Diner's Club card (I did not acquire a credit card until many years later). My studies ended and we traveled north together. I put my belongings in storage in Luxembourg and we went our separate ways. One day, after sleeping at the Brussels youth hostel, far from the center of the city, I turned a corner and there was my artistic brother heading my way. (As an aristocratic, ascot-wearing teenager, he naturally stayed in first class hotels.) I spent the next few summer weeks visiting Holland, Denmark, Norway, and Sweden, sleeping in my tent and getting very tired of the 40 degree cold and incessant rain. I escaped by hitchhiking to Lisbon in eight straight days. From Stockholm to the border, I took the train on which I met a young Australian girl. She had a ticket right through to Copenhagen so I told her that I would hitchhike and meet her at the train station. I got there first and waited. It was early morning and commuters poured off trains and came out of passageways, a sea of grey-suited brokers, clerks, and lawyers, hundreds of them at once. Directly in the middle of these folks was a Laplander in full orange and black regalia including his fancy hat. I would not have been surprised if his reindeer had come trotting along behind him. No one paid any attention.

The girl got off and we continued south. About a mile from the ocean, where we would get the ferry to Germany, we were walking along the side of a four lane superhighway. It was raining and in the distance under an overpass, I could see a man and a woman waiting for a ride north. I mentioned that the man looked like a friend, but that was obviously my overactive imagination. We walked on and the closer we got, the more I became convinced that I knew him. I started hopping around and gesticulating. I was very excited, because it was Doug, whom I had left months ago and thousands of miles away on the Red Sea. His friend turned out to be a man in a kilt! I never saw Doug again. My new Australian girlfriend wanted to go to Bielefeld; I wanted a less rainy environment, so after only a day of commitment, we parted ways. My trip was uneventful and I arrived in Lisbon with a wonderful and knowledgeable Portuguese man who saved our lives by avoiding a lunatic who came at us at 60 miles an hour in the wrong direction. I visited in Lisbon (where I paid 13 cents a night to pitch my tent), Madrid, Zaragoza, and lovely Barcelona, before heading north via the Pyrenees, with the magnificent road sinking, climbing, and switching back and forth. Lamentably, it was high summer season, and all sun-seeking Scandinavians had been on the Costa Brava and now they were going home. The highway out of Barcelona resembled a huge parking lot. So I sat off on the side and read the *Time* cover story on Chagall. Eventually, someone gave me a ride. I hitchhiked all the way to Luxembourg, my Icelandic Airlines flight back to New York, and Marvin and Gail's wedding, where I had to wear a tuxedo, for the first time in my life, I think.

I returned to Europe in 1967, in order to study in Florence. My interests lay in art and architectural history and fluency in Italian. I wandered around Florence but did not do any other traveling. After a while, I came home to New York. Four years later, upon completion of my doctorate, Terry and I flew to Reykjavik, where I studied at the university. I had been in Iceland before, and on one of these earlier trips had flown downward almost into Surtsey, the volcano that had recently emerged from the sea. The lava and steam were overwhelmingly amazing from the window of this acutely angling plane. Had the commercial pilot not pulled up we would have been vaporized. We hitchhiked from Reykjavik in the south all the way to Olafsfjurther on the northern coast from where we could see the island of Grimsey and the Arctic Circle, some 15 miles away.

Few cars passed, as we stood waiting for rides. One took us to a whaling station where we saw a comparatively puny sixty foot whale being flensed. (It is a great sadness that Iceland, Norway, and Japan continue to kill our sea-going brothers and sisters.) Next, along came a fancy Citroen. We got in and I began my long recitation: Do you speak, English, Deutsch, Italiano, Français, pah Russky? Apparently not, but I still do not believe it. All Icelanders study English, Danish, and even German. This was a very sophisticated and wealthy man. He was justifiably proud to be an Icelander and refused to meddle in other languages when he had his own sacred tongue, unchanged during the past millennium. As we flew through the lava fields, he would say something, much of which I did not understand: *Vartha*! I quickly looked it up in the dictionary, turned around, and sure enough, there was a large cairn receding behind us. I thanked him profusely, *Tak*, as we left and began another long wait, so long that we sat down in the Laxdael (the Valley of the Salmon, and the scene of the *Laxdael Saga*) in order to eat lunch, but, yikes, here came a car, which we were very loath to miss, so we threw the sandwiches into our packs and stuck out our thumbs. Somehow we managed to fit in a car overburdened with four suited businessmen. Many hours later, we arrived in Olafsfjurther and requested the camping ground. No, it was one AM and this local postmaster insisted that we sleep at his house. I countered, but lost. In the morning, his daughter took us, at breakneck speed, along the high and dangerous coast road, part way back to Akureyri, the northern capital. We thanked them too.

Eventually we arrived back in Reykjavik, got out of the big truck, and Terry picked up a stack of kronurs that someone had lost. We walked all the way back to our rented room on Ranagata, which had inspired Haldór Laxness to write *Atom Station*. Iceland is an expensive place to live and with money pouring out for eight dollar pineapples and $1.00 Campbell's soups (we bought neither), we left for France, where fresh fruit (other than bananas grown in green houses) was readily available. We worked our way around France and Germany and then returned to the US.

From February through May 1972, we drove around the periphery of almost the entire US, some 10,000 miles. Not once did we sleep in a motel, hotel, or lodge. We camped (in the cold), visited national parks, and stayed with friends. Our dangerous descent into the Grand Canyon on an abandoned mining trail and a walk across steaming Death Valley (159 degree ground temperature) were the high points of this venture. Mickey, Annice, Terry, and I drove from California back to Arizona, and proceeded to descend into the most amazing geologic entity on earth. The old mule trail was not used very often and so it was in bad shape and a careless step could lead to a fatal fall. Annice had never done this type of thing before and had a very hard time; Mickey had to carry her pack in addition to his own. When we reached the midpoint, we set up camp. I wandered off on a side trail to locate a spring; as I walked along, the trail got narrower and eventually it became so dangerous that I turned back. Had I fallen, no one would have known where to look for my body. I discovered a small tunnel into the mountain; at the entrance to this old mine, I found a shovel and an ore car. I doubt that anyone had seen these things in half a century. When I returned, Terry and I had a quick, simple supper, but Mickey and Annice constructed a circular enclosure out of small boulders and then cooked a fancy meal replete with herbs and spices. The next day, we arrived at a narrow beach where we camped. This was the only time in all of my travels, hikes, camping trips, and climbs that I drank alien water. I purified murky Colorado River water with iodine tablets; it tasted like medicine. We then reversed the process and climbed back out. Just a few hundred feet from the top, we encountered a straggling group of college students descending. Although I am very chary of giving strangers advice, I casually mentioned to a young woman, who was reading a book, that if she failed to pay attention to where she placed her feet, she would probably end up a fatality. The 17 hard miles consumed four full days. (I once walked almost 40 miles in a single day, part of which was across the slippery and crackling ice of Lake Constance.)

Two hitchhikers accompanied us across the desert. The Italian stripped off all of his clothing except a small pair of trunks; I explained that he would get badly sunburned but he did not listen. After a while, Terry said that she must paint (despite the intense heat), so we set up camp. The three men continued in order to locate a spring. As we walked, a jeep pulled up. The driver had a ca. 20 gallon tank of ice water and gave us some. The Italian and I returned to Terry, but the explorer continued. His father had been the original surveyor of the lowest point in Death Valley, which is also the US low point, and he felt that he had something to prove. He did not make it back that night despite the enormous fire that Terry and the other man built and their howling to attract his attention. He showed up in the morning; he had become confused and spent the cold night alone. Luckily, he had an emergency aluminum blanket with him. We walked out, thus completing a 26 mile round trip adventure.

After Fred and I climbed Orizaba for the first time, we drove to an infrequently visited archeological site. El Tajin comprises a group of exquisite excavated Totonaca temples and some earthen mounds under which, apparently, lie more architectural treasures. Except for one or two other Americans and some Mexican children selling rattles and little dirty bottles of an unknown liquid (which turned out to be pure vanilla), Fred, Judith, Terry, and I were the only visitors. It was a Saturday and so the *voladores* were setting up to fly. The men spent 30 or more minutes preparing their ropes, laboriously climbed the 100 foot pole to the tiny platform, and proceeded to jump off and fly around and around. It took 10 minutes to reach the ground. Four men flew and one, who was about 80 years old, beat a drum. Fred did not bother to watch.

Figure 9: El Tajin: Temple of the Niches, by Terry Hauptman

AN ADVENTUROUS LIFE

In 1993, Terry and I circumnavigated the globe. This was an amazing trip. The plane tickets cost $7200, more than necessary because we wanted to visit both Australia and New Zealand, which required a latitudinal shift. Local transportation, additional flights, food, and lodging brought the total cost to about $18,000. That is a lot of money for a frugal man. It was worth it. Fiji had been hit by a horrible storm and the bridges that would have allowed us to drive across the main island from Nadi to Suva, the capital, were both lying in the water. Still, we managed quite a bit. Each country offered its own pleasures: intense heat, Aboriginal culture, ascent of Ayers Rock (Uluru), the Great Barrier Reef, and many lovely urban areas in Australia; indescribably exciting and beautiful geologic activities, snow-encrusted mountains, Katherine Mansfield's house, and Maori culture in New Zealand; extraordinary architectural works all across Japan, even in the most humble private dwelling, Kyoto's 1400 temples, Tokyo and the Palace gardens, the horror of World War II Hiroshima, well-preserved for our edification, Mt Fuji; Hong Kong; China (by ferry, bus, and train); and many fascinating surprises in Tel Aviv and Jerusalem, for example, the Shrine of the Book, under construction when I was there in 1965, was finished and we could inspect some portions of the Dead Sea Scrolls, not long after passing the Qumran caves where they were found. And I was able to climb to the top of the mesa on which Masada is located. Everyone else on the bus took the gondola.

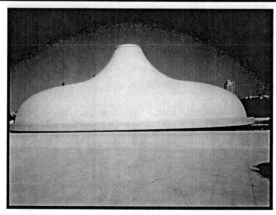

Figure 10: Maori Sculpture, Ayers Rock, Terry in Hiroshima, Shrine of the Book, by Terry and Robert Hauptman

I next turned my serious attention to visiting mountain sites, and Fred and I have taken countless trips around the US and Canada in order to climb and to do research for both *The Mountain Encyclopedia* and *Grasping for Heaven: Interviews with North American Mountaineers*. In some cases, we have returned to the same (esoteric) locations in the Cascades, Rockies, Sierras, Tetons, and other ranges over and over again. Sometimes we would drive very long distances (many thousands of miles) and so we had time for lengthy, desultory discussions on every conceivable topic: we covered religion, cosmology, physics, classical and popular music, literature, but also extremely personal beliefs and experiences. Fred thinks nothing of driving many miles to take a look at a geographic point, and one day, when Rainier was overwhelmed with fog, we hopped in the car for what I thought would be a short trip. We wanted to see the most northwestern point in the lower 48 states, an exquisite rocky cliff face with deep undercut caves created by the Pacific Ocean's pounding waves. We turned around and went back to our motel. The sojourn covered 500 miles. Fred enjoys controlling a car and is an excellent driver; he can go for 50 miles after an 18 hour climb, and he does most of the driving. I sleep (which is a bit unfair). Once, when I was driving across some barren stretch, Fred read Peter Hackett's book on altitude sickness out loud. Who knew that there were so many ways to die?

I did almost all of my extensive traveling on my own. Naturally, I accompanied the Wagner group wherever it went but once I graduated from college, I was much too independent to take tours or cruises. Nevertheless, I had four occasions to take a tour, when refusing would have meant not being able to go where I wanted to. First, came Ayers Rock in the middle of the Australian Outback. From Alice Springs, one must either travel the three hundred miles with a group, walk, or possibly rent a car. If the car breaks down, the nine species of virulent snakes, the kangaroos, or the blazing heat will certainly take their horrific toll. (Cars, usually Land Rovers, that venture out carry water, gasoline, and two extra spare tires on their roofs.) We chose a tour, which was very beneficial because the guide was extremely informative. All 50 or so people on the bus, including some who did not seem capable, climbed Ayers Rock, which takes some doing. (Because tourists were disrespectful and littered this Aboriginal holy site with graffiti, it is now closed to climbers.)

When we reached Hong Kong, I discovered that we could not just waltz up to the Chinese border and receive a visa. Had I known, I would have gotten one in Washington (which is what I did for Australia). I refused to surrender our passports for 72 hours, so we took a one day tour into southern China: Shenzhen, Guangzhou (Canton), and other places by ferry, bus, and train. We missed Beijing, Xian, the great Wall, and a few thousand other places, but we did the best we could. The third necessity was Masada. Only tour buses go out into the desert here, so again this was a good choice, and the guide was extremely knowledgeable. We passed the Qumran caves, which I would have missed had I rented a car. Finally, Fred and I wanted to get close to McKinley (Denali), but only landowners are allowed to drive inside the National Park boundaries. Our eleven hour bus ride, which stopped often to allow us to see Elk, Grizzlies, a rare black wolf, birds, fish, Mirror Lake, and the mountain, was well worth the cost. It is not necessary to be completely autonomous.

When one travels to distant locations with some frequency, many things occur, some positive (it is easy to help people in trouble) and some potentially fatal. The best experience occurred to someone else. When I was hitchhiking somewhere in Europe, I met a youngster who told me that two people were picked up by some Italian men. They not only drove the Americans to a destination and fed them, but also gave them what amounted to one hundred dollars. When questioned, they explained that they did not want people to think that all Italians are gangsters. Who stereotypes an entire nation in this manner? Apparently a lot of people because when hitching in Sweden a man picked us up and then proceeded to work himself into a real frenzy (he was quite angry) because he thought that everyone believes that all Swedish women are sexually promiscuous. And that reminds me of two kind Dutch drivers. The first, because he picked me up and we had a fine conversation despite the fact that he was very badly facially disfigured, perhaps from a war wound or fire. The second also gave me a ride but he had a secondary motive. As we passed Edam, visible off to the side of the highway, he invited me to come home with him. I always tried to avoid this because I did not want to take advantage (of "the kindness of strangers") but in this case I definitely could not accede, because he prefaced his invitation by saying, *Du bist ein schöner Junge* (you're a beautiful young boy). He was slightly deluded. I am not gay; I could not accommodate him. So I missed out on all of the good cheese he used to seduce me into visiting.

The negative experiences are more memorable. Once, when driving from Minnesota to Vermont alone (after Kira came into our lives, she and Terry flew and I drove), I stopped for the night in Gary, Indiana. This is a very dangerous area (motel clerks are protected by thick plate glass barriers). At five in the morning, I was awakened by a truly horrible argument next door; these people were shouting very loudly and would have frightened Clint Eastwood or Arnold Schwarzenegger. I did not wait for things to escalate out of control; I did not want to get shot or be called as a witness in some legal imbroglio. I was out of there so quickly, I probably forgot my shoes. (I was not really armed and had no unequivocal way of defending myself against a shotgun wielding sociopath.) The same type of thing occurred when Fred and I arrived very late at a motel, but it was our fault. I try to tiptoe around, not because I am scared of other guests (I am) but because I am considerate (and also expect these fools to be when I am trying to sleep at nine PM because we plan to arise at three in the morning and they are raucously partying). Fred is a bit more energetic and would never think of whispering or taking his shoes off (even though I sometimes alert him to the situation). I guess we must have really annoyed and provoked someone (undoubtedly the single most egregious asshole I have ever encountered) because at about six the following morning we were awakened by a blaring TV in a locked room. The motel owners were not even awake yet and it took a long time to extinguish the excruciating noise which bothered (punished) us but also everyone else in Shasta or Flagstaff or Jacksonville or wherever we happened to be.

ELEVEN
SKIING AND MOUNTAINEERING

Before I spent my 13th year in Vermont, I knew nothing about skiing. During that winter, we devoted our weekly Wednesday afternoon gym period to skiing at Bromley, one of the first such developments in the east. It was owned by Fred Pabst who also owned the Blue Ribbon beer. Skis, poles, and boots were provided but we had to climb the hill, though I may have ridden the J-bar once or twice. After I learned how to control myself, to a limited degree, I liked to climb high and come swooping down very quickly. And that is the type of skiing I always preferred: fast, direct, and with many jumps. Flying through the air is exhilarating. But once one has gone up and down 10,000 times at almost every major area in New England and then in Austria (including St Anton, Lech, and Zürs), it apparently becomes a chore. I got bored and for the most part stopped. (The increasing expense and inconvenience were probably also factors.) I shifted to cross country and this was a practical necessity when we lived in our little cabin, and for others too. One lovely winter day, we were inside reading, when there was a loud knock on the door. It was Eleanor Clark and a very well-known literary scholar. They had used cross-country skis to come to us from the back side of the mountain, where the Warren's had their house. They could only have walked, snow-shoed, or snowmobiled, since at that time the two miles of country lane were not plowed during the winter. All of this is now a distant memory.

Most people ski primarily or exclusively because it is physically exhilarating. I enjoyed this aspect very much, but even at a comparatively young age, I took great pleasure in the beauty of the snow-clad mountains. Most skiing in the east is done below tree line, so after a storm, the foliage is encrusted with a sometimes enormous layer of glistening snow and this can be enjoyed as one ascends through the woods on some conveyance. One early morning, I went high above Innsbruck and the cloud layer, and was treated to millions of little wafting snow crystals glistening in the new sun's rays. (The same is true for climbing, and mountaineers often cite esthetics when asked what compensates for all of the necessary suffering.)

In 1955, there was nothing more important to me. In the off-season, I read, I practiced kinetics imaginatively, I went to films, and I waited for snow. Once, when I was perhaps 17, I told my mother that if I could not ski, I did not want to live. (Well, we sometimes say stupid things.) Somehow, I learned that a small group of people had just founded the Staten Island Ski Club. I went to a meeting and joined. These were all young adults (20-30); I was 14 or 15. I am still stunned that they welcomed me (and one other youngster) and took us along in cars or rented buses, whenever we were able to go. I always went out of my way to be polite, courteous, and helpful, so they would continue to allow me to accompany them. One member bought new equipment and he sold me his old, overly-long hickory skis for a reasonable price. We left Friday evenings and drove for hours to Vermont's Mt Snow in the South, Killington in the center, Sugarbush farther north, and even Stowe, not all that far from the Canadian border. On Sunday afternoon or even later, we drove all the way back to the City and work or school early Monday morning. Skiing was very important to many of these folks, but since they were all young and of both genders, the trips also functioned as social interactions. Indeed, one loyal member had cerebral palsy and could barely walk; he certainly never skied, but he liked to come along nevertheless. Two were married but the others were single, and some dated each other (and eventually married). I sometimes felt like a third wheel, but I made myself useful. One day, when I was perhaps 16 or 17, Rudy was already extremely inebriated at breakfast, falling over onto his bacon and eggs. We were all on the bus ready to leave, for some reason, and I was asked to go back in and get him. (I was friendly with him and had done construction for his company.) I tried.

It is very difficult to convey extreme emotion, so the reader probably cannot fathom the extent of my commitment to this elitist sport but this may help: One day, at Mt Snow, I skied for perhaps eight straight hours. (I usually did not stop for more than a few moments to eat something.) We returned to our lodge and then, after rolling off this enormous mountain with its broad slopes and long trails all day, a few of us went night skiing on a tiny hill serviced by a diminutive rope tow. I could not get enough of this foolishness, but ironically, I never was very good at it.

Time passed and I lived in places where skiing was hardly an option so I cut back. But even recently in Minnesota, I would try to go a few times each season, sometimes with a friend but often alone. Now that I live near many Vermont ski areas, I almost never go. Starting when she was three, I have asked Kira many times if she would like to learn. I had planned to buy her a harness and connect her to my waist with a short rope. Now that she is eleven, she could learn on her own, but she has always refused. She lacks motivation, and I am not going to force her to take up a sport (skiing or snowboarding) that demands a thousand dollar equipment investment, and is, additionally, quite dangerous. Paradoxically, she is now participating in a sixth grade snowboarding program at school. When I skied, no one wore a helmet. Now helmets are seen frequently on youngsters flying, rolling, or tumbling down the fall line.

In 1971, in Reykjavik, I met someone who climbed and somehow managed to get him to take me along on an attempt of a small peak not far from the city. It was, naturally, a very nasty day, rainy and extremely windy. I protected myself with a large rain cape. It was impervious to water but the wind twisted it around like a sail. It was not very pleasant working my way up to the plateau where I found some snow and ptarmigans in their white winter feathers.

Many years passed, and in July of 1989, I was on my way to Reno to speak on a panel dealing with law and the provision of information. For some reason, we were briefly in Boston visiting. (Perhaps I was flying out of Logan, but I do not think so.) Judith had a new boyfriend and somehow we got to talking about mountains. Fred is a very serious mountaineer, but he had not climbed much since coming to MIT from the University of Paris some years before. He suggested that immediately upon my return, we should drive to central Mexico in order to climb Orizaba, at 18,800 feet, the third highest mountain in North America. I had been hiking, scrambling, and climbing (in an amateurish way) most of my life, but I lacked the knowledge, skill, and serious conditioning for this bizarre undertaking, which did not bother Fred at all. Somehow, he managed to convince me and off we went to Harvard Square to buy some guide books. I flew to Reno and back and along the way picked up my old French mountaineering pack and bought crampons and an ice ax. The trip was either the most exciting venture of our lives or a disaster. The evaluation depends on who is consulted. Terry was in heaven. I had a very hard time with the excessive heat, the fast food eating establishments with their hamburgers and bacon, and Fred, who did all of the planning and most of the driving. Much to my utter amazement, he had the route precisely planned out (with AAA marked maps) and reservations all along the way. Never in my life, in my tens of thousands of travel miles in what would be some 40 countries had I ever reserved a room (except when attending conferences); I never knew where I would be. Thus, when we arrived at the end of the first day's travel with the sun still brightly shining, and were I alone at about the midpoint of what I would have accomplished, Fred pulled off the highway in order to locate a motel. Well, there were about ten of these establishments in close proximity to the exit ramp, but he drove off into the Virginia countryside. I was most confused and asked what was going on. He was looking for Motel Six or Eight at which he had reserved rooms and it was not in evidence. After a few days of this madness, I got used to his way of doing things, and that is still how we operate.

There is an advantage to this; I have often been forced to scramble from one motel to another when all rooms are booked. It is most disconcerting after driving alone for 16 hours and I am in need of a restroom and sleep. When traveling by myself and with no precise destination, I **never** make reservations. When Terry and I flew around the world, we never knew where we were going to be, and we often drove or walked arbitrarily until we located something (in Fiji, Brisbane, Tokyo, Hong Kong). Just a few years ago, I drove by myself for many weeks through all four American deserts, thousands of miles, in order to take photographs for a book, and trusted to luck, even in Death Valley, to find something. I always did. I have even flown without reservations; buying the ticket at the counter is hazardous and very expensive, but if one has frequent flyer miles it can be free. It was!

We finally reached Tlachachuca at 8,000 feet. Terry and Judith drove off to Mexico City and Fred and I began to climb, across the corn fields and up the steep slopes. We went from sea level at seven AM to almost 14,000 feet by eight that evening, when we hit an un-crossable abyss. Fred felt the effects of altitude; I was okay. It rained and we suffered under a stretched tarpaulin. The morning brought more rain and fog. We started down, never having located the refuge. After a while we came to a settlement, noted on the map at 12,000 feet. The little trail went right though its garbage dump, where naked children cavorted. We kept moving and after many hours arrived back at our starting point in the city. Terry and Judith picked us up and we started for home. But it was too hot and the plastic radiator fan melted. We lost many days attempting to get it repaired. Fred had an important meeting to attend in Boston, so when we finally reached Houston, he flew home and I drove all the way back to New York. I did not make any reservations! The climbing, as it turned out, was inconsequential, but Fred did compliment me on my strength and hiking ability. That was heartening for Fred was 28 and I was close to 50.

Figure 11: Bob on Castle Peak, CO, Along the Ridge, and Just Below the Summit, by Frederic Hartemann

Figure 12: "Bob on Columbia Ice Field", by Frederic Hartemann

Figure 13: Bob on Mount Lassen, by Frederic Hartemann

A few years later, Fred was in France working at Thompson, the enormous electronics company. I was on sabbatical and so I flew to Paris, and off we went to Chamonix to climb. We did the Aiguille du Tour but turned back in a whiteout. We also tried the Gran Paradiso but the snow was so deep, we barely got out of the parking lot. About a year later, Fred was back, at UCLA, and he visited in Minnesota. We drove north to a casino, but immediately lost interest in gambling. We returned to the car and Fred took a look at a map and noted that the Minnesota high point was just a few hours away (near the Canadian border), and so we drove north, located it, and climbed. Thus began our quest to reach the high points in all 50 states. We persevered, in trip after trip, climbing successfully or failing and returning again (and again). It took me nine tries on nine separate occasions to reach the summit of Mt Whitney, the highest point in the lower 48. This is neither a technical climb nor very difficult, but it is quite long (22 round trip miles with a 6,000 foot vertical rise) and I ran into many different problems (altitude sickness, loss of energy, fear, error) on both one day and two day attempts. (Success came on a 16 hour, one day extravaganza.) Fred made it to the top five times. Some of our trips were quite extensive, hitting many state peaks: Oklahoma, Arizona, New Mexico, and Kansas on one and many East coast states from Maine to Georgia on another. Over a 14 year period, I managed to stand on the summits of 45 high points. Then I stopped. I still have Montana, Wyoming, and Washington plus Hawaii and Alaska to go. Interestingly, I have climbed two of these but failed to reach the summits and I have been to all but Wyoming's Gannet Peak. Fred made it almost to this mountain, a 26 mile hike, but then turned back. Fred has about 42 peaks including Mauna Kea and Washington's Rainier. We have done many other climbs as well and never used a guide. And at one point, we returned to Orizaba and tried again, and at the correct time of year. Fred reached about 17,000 feet. I did not really leave the 14,000 foot refuge. I continue to climb year-round in Vermont where the mountains are smaller, but challenging enough, especially in winter. Sometimes Kira, at seven and eight, has accompanied me. She struggled but made it to the top.

Figure 14: Bob on a Steep Slope, by Frederic Hartemann

Struggle is inherent in mountaineering. The things that normal human beings must do or bear in order to reach high Himalayan summits are incomprehensible: days spent trapped in flimsy tents, 24 or 48 hours of continuous work and often with nothing to drink at an altitude where dehydration occurs even when consuming liquid, intolerable cold that destroys cells and fingers and toes. It is a very hard life. It would be much easier and safer to take a tow or gondola up and schuss back down on parabolic skis. But we prefer not to. One day, Fred and I arrived at Mt Katahdin, the highest point in Maine, the terminus of the Appalachian Trail, and the hardest mountain east of the Mississippi. The stolid guard would not allow us into the park. She does not care about people; her concern is her tiny quota. Fred, who grew up in Paris, is very volatile. He exploded. I quickly drove away and listened to him rant for at least an hour. We went home. We returned a year or two later. This time, we arrived at five AM, so we got in but there were stipulations: Do not climb the mountain; if you do, do not go above tree line. We paid no attention. We started up in the horrific rain and wind. Fred announced that we were to do the knife edge; I countered that I would not. I weigh about 130 pounds. The wind would blow me off this precipitous, narrow ridge (as it blows people's gloves, water bottles, and packs off mountains with some frequency). I would very much like to walk this wonderful mile but not at the risk of my life. We were both upset and did not talk. When we reached the midway point, where a little building offers some shelter, we stood looking at the high cliffs. And then Fred was gone. This will seem impossible to readers but since it has occurred to me a number of times, it is a real manifestation: one minute Fred is right there, the next he has disappeared (as on Whitney once when I thought he had fallen down thousands of feet, and I turned back). So I decided that he had headed up the trail and I ran after him calling his name. Because of the bad weather, no one was above me on the entire mountain. I raced along but never found Fred, because he had simply gone off to a restroom. For many hours, I continued upward. Finally, I reached a steep slope and thought that I would just climb up to the plateau and take a look around, but when I got up, I could not stop moving and continued from one large cairn to another until I reached the summit, where the fog was so thick I could hardly see the big sign. I carefully turned back to the correct trail and descended. Near nightfall, I reached the parking lot, and you may well imagine that Fred was incensed. He had had to wait on a porch, but still in the cold rain for many hours, since I had the car key. He had thought that I had gone down. We worked it out. Fred had once climbed Katahdin in winter cold (50 below zero) so we both now had another highpoint. I subsequently climbed it again; the weather was equally horrendous and it took ten long hours!

Mt Hood is Oregon's highest peak; it is also one of the most frequently climbed mountains in the US, both because it is only about 11,000 feet high and also because it is not far from Portland. For these reasons it is also the site of many horrific tragedies. We foolishly left Salem, at sea level, at midnight after only four hours of sleep. Unacclimatized, we headed up at three AM. The entire climb covers only three miles, but it can be confusing and dangerous as one approaches the summit. Because frozen rocks are released and plummet down as the sun warms the terrain, it is imperative to get down off the summit before noon. We did, almost exhausted in my case. A week after a subsequent

Hood climb, I was driving from Minnesota to Vermont when my cell phone jingled. Terry's mom informed me that people had tumbled down from the Hood summit, hit others, knocked them into the enormous *bergschrund* just below the summit slope, and some had died. A rescuing helicopter had crashed. On this climb, in a whiteout, I was overly cautious and insisted that we turn back at the top of the climber's trail. We did. Turning back allows one to try again on another day!

Figure 15: Mount Hood Hogback with Climbers, by Robert Hauptman

And then, when the snow is off the meadows and the sun is beating down, one may fully appreciate the animals, fish, birds, and especially the flowers. Justice William O. Douglas, in *Of Men and Mountains*, describes, with some frequency, the many kinds of flowering plants he observed and collected over a long life-time of enjoying nature. He emphasizes both the countless species and the unimaginable profusion one finds once away from proximate civilization: In a single three foot square area he once counted 30 different species. Visitors to popular tourist areas (Death valley and other deserts, Rainier or Adams in the Cascades, the open slopes of New England mountains) at just the right times, will be met by vast carpets of varying species, the reds, pinks, yellows, whites, purples, and even blues flitting from one to the other or merging in a rainbow of ecstasy. One day, coming off the Teton range below Garnet Canyon, as we worked our way down the switchbacking trail, there, along a very steep, wide slope encompassing many acres, I was met by a sea of wild, yellow flowers, millions of them wafting in the breeze, a luxuriant carpet that was obviously invisible when we ascended long before the sun rose that morning. Of course, I am interested in identifying all animal and botanical species and I do so by using guide books and making inquiry when possible. But I do not wish to burden myself, as Edward Hoagland often does, by fetishizing identification to the detriment of immediate pleasure (which is why I almost never carry a camera). Seeing yet another bird species is less important than enjoying what is before me. I do not fly off to Kentucky because some fanatic has spotted an arctic tern; indeed, I do not even bother with a life-list any more. My knowledge and skills are most honed in areas where I have lived: I know animals, trees, and flowers in the Midwest and New England. The south and west present major obstacles because I have not spent enough time in these places carefully observing. The inhabitants of the Hawaiian rain forests and the Alaskan tundra are even more alien.

TWELVE
REPRESENTATIVE CLIMBS

Here I would like to offer detailed overviews of some often difficult but compelling climbs. Naturally, they are similar in many respects but they also differ, sometimes dramatically because of varying difficulty, extent, expended time, and topographical and climatological conditions. For many mountaineers, physiological reaction is perhaps the single most important factor; luckily, I usually acclimate quickly and successfully. I do, however, know how one feels when this fails because on my first Whitney climb, as I approached Trail Camp (at 12,000 feet), I asked a ranger how far away it was. She said about a third of a mile. I never made it because I felt so ill that I had to lie down on a rock. I then descended alone (sending a message to Fred, who was ahead of me, to let him know); this was the right thing to do. Headache, stomach pain, nausea, dizziness, and a general horrible malaise are all possible reactions.. If things progress, pulmonary or cerebral edema (followed by death) may occur. One would hardly think that such extreme reactions are possible at 12,000 feet, but on another ascent at the same location, a ranger asked me to climb up a bit higher and retrieve a backpack for her; the day before, a climber had become ill, dropped his enormous, new, and expensive pack and raced down to 10,000 feet, where he was evacuated by helicopter. No one does this unless he or she is extremely incapacitated: It is embarrassing and potentially very costly.

Mount Mansfield

I went to Colorado to interview two mountaineers and to climb. Despite the 13,000 foot altitude, lack of protection above tree line, and continuous 30 to 40 mile per hour jet stream wind that made communication almost impossible, the climbs were easier than on Mt. Mansfield, at 4393 feet, Vermont's highest peak. Between late November 2006 and late April 2007, I did six winter climbs along the Appalachian Trail of this ostensibly easy mountain. (The vertical rise is 3,000 feet in just over two miles.)

My first five climbs were on snow and ice from the parking lot to high points. On the sixth, I hit one fifty-foot stretch of open ground low on the mountain and some rock along the summit ridge; otherwise, the mountain was replete with deep snow. Diminishing sunlight, unclimbable ice, emotional strain, and extremely deep snow resulted in failures on the first five climbs.

In late April, I try again. The weather is spectacular, as predicted. It must be 80 degrees, much too hot for me. I climb and descend in a T-shirt, the sun burns my face and arms. I could have used some sunglasses. Despite the heat, the snow is excellent and only occasionally do I sink beneath the surface. I move upward on the trail, though as I get higher, the blazes get lower. In some cases, the blazes, which are at about five feet, are beneath the snow.

The snow around Taft Lodge is unbroken; instead the track I'm following goes through the woods, away from the normal route. I persevere and after a while come back to the correct route and then out into the open where the tree line begins to degenerate. The footprints in the snow go their own way, but they clearly lead to the summit, so I follow. I mark my path. I arrive at the base of the ridge. The route leads up a steep slope. I move upward carefully; a mistake can be painful. I reach the top of the ridge and look down. I have covered about 200 horizontal feet and perhaps 75 vertical feet; it is very steep in places. I begin again. This time I move up about 300 horizontal feet and 150 vertical feet; here the slope is sometimes 80 degrees and the snow is either extremely deep or, conversely, a scattering on the rock face. This is bad because I cannot get a good purchase with my ice ax. If I fall, I will tumble (I once did), take some snow with me in a minor avalanche, and perhaps hit rocks. I am unhappy, but I continue. Near the top, I turn around and ask myself how I plan to get down. (The ascent is less stressful and safer than the descent on extremely steep slopes; indeed, in big, serious climbs it is sometimes impossible to descend by the same route) But I know what I plan to do. I stride across the summit ridge, touch the benchmark with my boot and hand and continue without pause toward the microwave towers.

I stroll along, meet some skiers who have come up from the gondola, and continue, searching for the point where I must drop down to the top of the lift some 600 vertical feet below. I miss it and circle around the cliffs, coming to a steep slope. I know this is wrong so I turn around. As I come back past the cliffs, I see a short glissade track and some footprints. I'm not certain that they were made by someone climbing up, but I hope that this is the case, so down I go. Within a few minutes, I realize that this is correct. I make excellent progress and reach the lift in only 15 minutes. I look back up and am staggered by the height, sheerness, and beauty of the cliffs, which I had briefly contemplated down climbing. This would have been foolish, since they are very steep and wet and I am encumbered with crampons, snowshoes, a three-foot ice ax, and a disbalancing pack. My climb and traverse are complete. I hop on the lift and step off a few minutes later. I walk back to my car, which is just five minutes away. It took me five hours to do this lovely Appalachian Trail winter climb of an unrelentingly tough mountain (Hauptman, "Challenges").

Mount Whitney

As indicated above, during the course of the 1990s and later, I attempted Mt Whitney, at 14,495', the highest mountain in the lower 48 states, nine times. I have come close to the summit on both one and two day climbs. The latter allows one to sleep after climbing from about 8500 to ca. 10,000' over the course of about five miles or a little further (12,000' and seven miles, which is where we always stopped). But for some reason, I succeeded in a one day push of 6,000 vertical feet in eleven miles, which means that we covered a total of 22 miles that day. Walking this far on flat ground is not particularly onerous (I walked almost 40 once and part of it was on ice, you will recall), but this immense distance climbing at altitude is wearying for all but the most conditioned triathlete. (I once encountered a big, powerful man sitting on a rock just below Bierstadt's 14,000 foot summit, trying to catch his breath.) Naturally, we have had many adventures on Whitney (where we always took the normal rather than the shorter mountaineers' route or the technical climb). Once, when Fred Andre was along, we followed our own course through the snow cutting the lower invisible switchbacks and then ascending up a steep narrow defile. Another time, I set a fast pace and ran out of steam. And once, I thought that Fred had plummeted down thousands of feet, so I turned back. (He was fine.) Many times, we climbed the final half-mile slope that leads to the summit ridge, front-pointing in the steep, ubiquitous snow. But one day, I noticed that the famous 96 switchbacks that are set off to the left of the slope were at least partially visible, so I thought that I would try them, just for fun. Fred took the slope. Slowly we diverged so that we were no longer in solid communication with each other. (It is difficult and also embarrassing to scream back and forth.) I did well, even after the trail became snowy and sometimes extremely icy on precipitous portions, which are protected by steel cables. Just below the ridge, I went straight up a steep slope, protecting myself with very deep thrusts of my long ice ax at each step.

I got to the top, turned right and moved along the ridge attempting to get to Fred on the big slope just a few hundred feet away. But I was stopped by an enormous pinnacle, which I could not bypass. I looked down the slope I had ascended and could see no way to pass around. I then looked down on the western side of the Sierras to the Pacific Crest Trail and also was stymied. I tried to go around on this side by clinging to the precipitous rocks but failed and backed off because of the danger of a serious fall. So I decided to retreat, that is, to go back down as I had come up. But I could not locate my route, and the descent was too steep, so I thought that I would wait to be rescued. While waiting, I walked along searching for a solution and eventually I spotted a point at which the upper trail showed through the snow; I thought that I could safely descend to this. I did and soon made it to the top of the normal slope and the three mile ridge that brings one to the summit. It is here, by the way, that one finds a little bronze marker that informs the reader that he or she must get off the ridge immediately if the skin tingles, since lightning strikes are very common. (Once, upon descent, Fred mentioned that my hair had been standing straight up! On another occasion, we had set up the tent at Trail Camp and were hunching down on the balls of our feet, after Fred had tossed our axes as far as he could because of lightning, when a group of people passed by. They had been caught on the ridge and one woman was still fearfully trembling and on the verge of tears.) I hurried along. But soon thereafter, here was Fred coming back from a successful summit. I was still feeling fine and could easily have made the three mile trudge, and then the eleven mile return to Whitney Portal and our car, but I did not want to travel alone, so I accompanied Fred back down, sadly missing an easy opportunity to summit. One-day Whitney climbs take between 13 and 16 straight hours of hard work.

The most expensive of all of our many US, Canadian, Mexican, and European climbing trips also occurred at Whitney: After sleeping at Trail camp, climbing, and then returning to the Portal, we discovered that a bear had twisted the car door, broken the window, and stolen our fig newtons. We learned an important lesson: Never leave food in the car! The new door cost $1500.

Aiguille du Tour

I am here in France to climb, but despite the positive weather report, because I am nervous, it is an inauspicious early Sunday morning for me. We prepare our equipment and drive to the little town outside Chamonix from which we will begin our ascent. We don heavy packs and begin trudging up the trail. The altitude is 1480 meters (ca. 5000 feet), and the trail is steep and narrow with abrupt drops on either side. After an hour or so, we come to the top of the first ski lift. Here, the trail swings away in a wide arc, the valley opens up broadly and minutely below, and towering mountains and glaciers ring around us in every direction. The trail turns into a small ridge, falling away, extremely steeply on one side. I wonder what would happen if I stumbled. With my heavy pack, I would not be able to catch myself and would tumble down the slope to my death. I am tired, the air is thinner, and I need a drink. But Frederic almost never stops (except to take an occasional photograph).

We are into the mountain now, which hangs steeply above us to the left and falls away violently, tumultuously to the right. One misstep on my part and Frederic, who is leading, would never locate my body. We encounter patches of snow here and there, which adds traction to the narrow trail. We move slowly but continuously. The snow increases where little avalanches have inundated the trail. What, I wonder would occur, if the snow came pouring down just as we pass? These questions constantly worry me, but every other climber seems to simply ignore them. The trail is so steep and difficult that iron bars have been let into the mountainside. I grip each bar and let myself down, cursing the cumbersome pack with all of the necessary paraphernalia that I will use on the glacier the next day. Again I see a bar, only this time the trail is virtually nonexistent and the exposure is frighteningly steep. I call to Frederic, who is well below this obstacle: How can I pass? He demonstrates some esoteric maneuver, with an equally obscure name. It looks impossible, but he urges me on. I manage to descend to him but wonder how I will ever be able to get back up on the return trip.

Evening is approaching and, naturally, I worry about not reaching the refuge before dark. We pull away from the mountain and out onto the glacial moraine. There, we walk along a narrow ridge of stones, each side falling away gently so that I no longer have to concentrate on every nuance of each step. After some three hours, I am almost exhausted. I look up and there, way above me, is the refuge, perched on the pinnacle of a sheer cliff. I wonder how it is possible to get up the steep slope that appears below the cliff. Naturally, the cliff seems impossible. Yet, as we continue, the ridge transforms into a steep incline littered with rocks and boulders. We move slowly, following red smears of paint. We reach the cliff, and I see a path alongside it. I trudge the last few steps, circle around the summer refuge (which is closed now) and finally find the path to the much smaller winter hut. Without my pack, I go back outside and look at the mountains and glaciers around me. We are now, after four hours, at 2700 meters (ca. 9,000 feet). My eyes fill with tears for the trip and the glorious, incommunicable beauty that surrounds me. Snow is falling and the wind is jesting with me. Frederic and I are alone in a small refuge that can sleep 30 cramped people. I continue to gaze down and there, far below on the moraine, is a lone climber moving steadily upward. I let Frederic know.

We relax and eat. Miraculously, there is gas for cooking, coal for heating, and even tidbits of food and drink. We wait. Lindsay arrives. He is a large, powerful man, a former guide in Nepal. He likes to climb by himself. It is now just about dark and it appears that we will be alone for the night. But suddenly there is a clamor at the door and in come two more climbers. Jenny and Richard have been out on the glacier and have decided to sleep in the refuge before venturing across the ice and onward into Switzerland. Their equipment is awesome: Richard has two lengthy and heavy climbing ropes wound around his torso. His pack probably weighs 60 pounds. They undo all of the equipment, make supper, and eat.

I say good night and go up to one of the three bedrooms, unfurl my sleeping bag (a weighty luxury upon which I insist) and try to sleep. I cannot. But just when I fall into a deep, comforting slumber (despite the 35 degree temperature in the room, the insanely howling wind, and the loud explosions of seracs breaking off from the glacier), Frederic informs me that it is five AM and time to get going. We dress, eat, and move out. It is 6:00 AM, snowing, and the wind is devastating. I question the sanity of continuing but everyone ignores me.

This time, I leave my pack behind, but carry my crampons and ice ax, and wear my harness. The first mile or so is over steep rocks. My task is more difficult because I am carrying things and cannot effectively use my hands. The air is thinner, so the work is harder. I struggle, annoyed by the encumbrances. At last, I see below me the beginnings of the glacier; the others are standing there roping up. Frederic points out the new snow. This is good, since we will not need crampons, but it is bad because someone must break trail in two feet of new powder. Richard and Jenny rope together, as do Frederic and I. Lindsay is un-roped. We head out, up the glacial slopes, steep and awesome. The wind is howling and crevasses loom on every side. We take a route that circumvents the larger ones and provides bridges across smaller fissures. We move slowly. It is tough going for me and I am not breaking trail. Hours pass, and the wind's intensity increases, obliterating our footprints and frequently causing whiteouts. I am worried: Will we be able to find our way back? Will I have the energy to get all the way down the mountain by nightfall? At 3300 meters (ca. 11,000 feet), and just a few hundred meters from the summit, I insist that it is time to turn around. Frederic wants to continue, but Lindsay agrees with me. Frederic is quite unhappy but relents.

Incredibly, Richard and Jenny decide to continue all the way across and then down into Switzerland. Since the route is unmarked and visibility is near zero, this appears to me to be sheer madness. Compass readings are fine in the desert but almost useless here, because crevasses can present impediments that necessitate extensive circumnavigations. We part company. Frederic flies down the glacier. Sometimes, I rest until the rope between us has played out and then quickly follow, running down slopes, digging my heels in until I catch him. Lindsay moves at a strong, steady pace placing himself at various points in relation to Frederic and me. Here he is, standing next to me, saying, *Let the rope play out because Frederic is crossing a hidden crevasse.* As he speaks, he falls in. I grab at his shoulder and help pull him out, thinking that I am a hero, but knowing that he could have gotten out without my help. We press on and finally arrive back at the refuge. It is nine AM.

The weather continues to roar. I wonder about the long trip back down the mountain. Frederic and Lindsay eat vast quantities of food, but I am not at all hungry. They insist that I force some food down so that I will be fortified for the two to three hour trudge back, one that will allow no breaks. We pack our gear and head down. Frederic is off, as usual, like lightning. He pulls away. Lindsay takes his time, staying close to me, offering encouragement. I am fine on the moraine and ridge because they are not especially dangerous. But next comes the tight trail with slopes falling impossibly away. We reach the single bar, the place that I found so difficult to descend. Now there is snow on the trail and no place to put my boots. It is straight up, perhaps ten feet. Lindsay indicates some crags and handholds. He tells me what to do. It is daunting but I do what he says and up I go.

It is nasty, but as we descend it gets warmer. I take off my gloves and hat. It is snowing and the wind is blowing mightily. The snow turns to a horrible heavy rain, and suddenly we are soaked, despite our protective clothing. We walk on and Frederic pulls farther away. Now that we have passed the difficult parts, so does Lindsay. Since I am alone, I rest now and then. The gap between us widens. Far below, I see the top of the first ski lift; it is at least a mile away and across the dangerous ridge I encountered when ascending. I worry that the wind will pick me up and roll me down the 80 degree slope. I hunch down and stick to the upper part of the trail. I am overjoyed to circle around and head down to the lift. My friends have long ago disappeared, and I am certain that they are moving down the last part of the trail to the valley. I am soaked and stop to rest. I lean against a boulder, and suddenly there is Frederic, emerged from the lift building calling to me. Hurry up, he shouts. My body is demolished, but I realize immediately that this means that we have a ride down to the bottom. I hustle to the building, step into a gondola, and fly though the air all the way down to the valley and respite.

That evening, we are sitting in a Chamonix restaurant. I happen to glance out the window and much to my utter amazement I see Richard and Jenny walking by. I jump up and rush outside, wanting to embrace them, for I feared that they were still wandering around in an endless whiteout high above us. I invite them in and we all have that French culinary specialty, pizza. They had given up their quest to reach Switzerland and returned to Chamonix just as we had. A few days later, Frederic goes up on the Mer de Glace, but I decide to take the lifts to the top of the Aiguille du Midi, where I once again run into Richard and Jenny, who had been climbing. During the course of the next few years, we make tentative plans to climb Mckinley together, though this never works out (Hauptman, "Beaten").

THIRTEEN
FRIENDS

My brother tells me that he has 600 names in his computer address book. Are these friends? It would be impossible to have even one tenth this number of real friends. Anyone who claims to have more than a dozen true friends is deluded. It is simply impossible to spread oneself so thin, commit to so many people in a meaningful way. Even parents will have a difficult time dealing fairly and lovingly with ten or twelve children. A real friend is someone to whom one is fully committed, for whom one may sacrifice, and who reciprocates. If one is not willing to make sacrifices, then there is no difference between a friend and an acquaintance or a stranger; ironically, sometimes we make extreme sacrifices for people we do not know. Real friends are not jettisoned because of petty, foolish, unacceptable, ideational, social, societal, or ideological differences. If one has, for example, libertarian beliefs in common with an ostensible friend and the person makes a radical shift and decides that socialism is superior, one does not write her off; if one does she was no friend. This, of course, is similar to spouses who love each other, raise a family, and stay together despite extreme differences in religious commitment. (True love is not extinguished because of dalliance, which is easy to forgive. It is the social taboo that has been drummed into people that makes it appear as if this is unforgivable.) Real friendship neither diminishes nor extinguishes ("Love is not love/ Which alters when it alteration finds..."). The commitment burns on despite circumstance or intervening distance. Consider an extreme case: A 50 year friend elopes with one's 18 year old daughter; this would upset most normal people. But although it is anomalous and peculiar and perhaps upsetting, if the two lovers make this perhaps foolish choice, a friend will affirm it. If a 50 year old friend harms one's innocent child, one must conclude that he or she is no friend at all. Friends remain available to help and succor in the same way that one would react to a long lost parent or sibling. Indeed, real friendship is similar to the bond that exists between parents and children or siblings and has little to do with the superficial "friendships" to which people frequently refer, and which come and more easily go.

Real friendship does not recognize boundaries. This, of course, is especially the case when one is befriended in a critical situation such as a war, natural disaster, or epidemic. When someone helps or saves a life, friendship is solidified and usually reciprocated. That is why (and how) college fraternities (and sororities) succeed. Humans are social creatures and, when newly arrived on a traditional campus, often need support, even in today's coddling environment, where there is an office or group for every conceivable national, ethnic, racial, linguistic, sexual, or physical orientation. People join these real (rather than virtual) social groups for real companionship. The inductees (pledges) learn that they may have to suffer a bit to become full-fledged members, just as their predecessors did, but then they will make friends for life. And I am sure that this does occur, but people change and forty years later, these youngsters probably have forgotten the good times they think they had and pay no more attention to their brothers and sisters than they would to a former but long forgotten fellow ball player or cheerleader. Friendships are not manufactured in social milieus nor constructed out of similar interests. As a highly eccentric and individualistic thinker and actor, I, naturally, never even thought of joining a fraternity (or even an invitational honor society, or most other groups or organization for that matter), but I did note how my classmates evolved. Their close fraternal relationships often dissipated, degenerated, and atrophied; they made new friends, some of whom remain steadfast, filling the lacuna left by their missing brothers and sisters.

Naturally, during my elementary and high school years, I did things with others both at school and with children who lived in my neighborhoods. But these were not friends and I would have said so even when I was 10 or 15. I had a delimited social life, though I certainly did many sometimes exciting and dangerous things. During my thirteenth year, which I spent with the Melendys in VT, I was close to Harold, but after I left, I never saw him again, and now it is too late. In college, I had roommates, and interacted with others (talked, wrestled, played basketball) but here friendship was elusive.

At one time or another, I have had perhaps fifteen true friends. One passed away at a very early age. Two betrayed me, which leads me to believe that they were not friends at all. Three were lovers whose long and full commitments preclude eliminating them merely because the relationships occurred 20, 30, or 40 years ago. Four are very close to Terry and so I share this commitment. Five are people with whom I enjoyed meaningful life experiences for extended periods of time. It does not matter that I am no longer in contact with R; it is irrelevant that I see Carol every week or two, Hugh and Jeanne once a week during the summer, Atida infrequently, and others only every few years. Indeed, sometimes years go by before I get to see my brother again, and he lives just a few hundred miles away. He is still my brother despite the infrequent personal meetings; kinship and commitment remain constant.

My lovers were my friends. I cared about them and helped and succored as I could, not always and not in every case but often enough and especially in long term relationships, and this despite the eccentric, anomalous, asocial aspect of some of these interactions. Even when broad revelation would have had at least some untoward consequences for everyone concerned, I never went out of my way to be secretive or dissimulating. I stay in contact with these people, when I can. That's what friends do. To remarry within a strict, orthodox sect and to inform a "friend" that she should **never** call again, as someone once told my mother, is indicative of a lack of commitment where it truly matters and the substitution of a false sense of loyalty (out of fear) to a new and demanding fanatical patriarch.

In order to celebrate our marriage, Terry and I climbed Stratton Mountain. By 1968, I had already climbed this southern VT landmark many times but this was Terry's first ordeal. She enjoyed the long trip through the forest with its many stands of conifers (eastern white pine, hemlock, spruce) and hard woods (paper birch, quaking aspen, ash, maple, and especially the enormous and impressive yellow birch which grow across much of the mountain). When we reached the summit we discovered that two new fire lookouts were in attendance. We had much in common and they allowed us to set up our tent next to the fire tower. We still talk about the lightning storm that followed, because we could easily have been struck (whereas Hugh and Jeanne lived in the cozy, tiny cabin they still seasonally inhabit 44 years later!). They would return to Buffalo during the winter or reside in our little town or live in New Jersey or Greenwich Village, but we remained very close regardless of where they might have been. We too went off to various places including Ohio, Oklahoma, and Minnesota for at least part of each year, but we always reunite during the summer and they often have stayed with us when they come down off the mountain for their weekly resupply trips.

One day, 40 years ago, I used my lunch hour to drive to a little nearby town in order to visit a body shop; I thought that I might have our Volkswagen repainted. On the way, I picked up a hitchhiker. Bob eccentrically refused to drive and so he walked vast distances in the rural Vermont countryside to get wherever he had to go. This was never an impediment to D. H. Lawrence's characters, who would walk five miles from the train station at midnight upon returning from partying in London, but pampered Americans prefer their Fords and BMWs. Bob was different. Indeed, he had never gone to college but somehow we immediately began to discuss literature and especially poetry. This too is extremely anomalous. Most loggers and stone masons do not care much for Eliot or Berryman. Bob does. And so began a long and fruitful relationship because of which we exchanged weekly letters for many decades. Eventually, I sold Bob's as well as others I had received from many well-known people to Yale University in order to protect them and make them available to researchers, despite the fact that some contained Bob's harsh criticisms. In fact, it is because of these acerbic remarks, and despite the many hours we spent together building, that I finally stopped communicating (although I recently recommenced sending material through the mails).

Bob's enormous literary and cinematic knowledge always gave us much to discuss. Actually, it was more fun to test each other on the most esoteric literary trivia (name four works with black in their titles: *Black Rain, Black Swan,* and *Black Orpheus.* I cannot think of another, but Bob, I am sure, could).

Marvin and I were thrown together at birth by circumstance and we have remained friends for 70 years. We do things differently, have led very different lives, and have different attitudes and beliefs, but this apparently does not matter. Friendship is not predicated or based on religious, ideological, or political beliefs, social standing, wealth, or other tangential matters. (In fact, I have never discussed politics with him, and so am not certain that he is a liberal or conservative, radical or reactionary; I honestly have no idea.) When we were young children, we spent a great deal of time together generally in his house; occasionally, I even went off to the country with Marvin and his parents. When at 12, I moved to Staten Island, I would return to Manhattan on weekends, often by bicycle, and stay with him. One evening, Marvin went out and I stayed behind. His dad, Charlie, arose at four am (and later much earlier) so that he went to bed at eight. At about ten, I was probably watching television, when Charlie got up and asked after his son. I replied that he had gone to Chinatown to shoot pool. Charlie was incensed, got dressed, and off we went up Catherine Street to search, although how we were supposed to know precisely where he was remains a mystery. Amazingly, as we walked north, Marvin came strolling south. I suppose that by the time one reaches 17 or 18, he can no longer be put in time-out or grounded. Still, Charlie was really upset. Years later, I went to the hospital a few hours after Marvin's first child, Jody, was born, and when Marvin and Gail moved to Long Island, I would often visit at their new home.

Friendship is not dependent on contiguity. Derora was a charismatic woman beloved by everyone she encountered. She and Mickey lived in a lovely country home in Athens as we all pursued various degrees. She is still my friend, although she succumbed to diabetic shock and passed away at a very early age. Garret now lives in Laos and only returns to the US every year or two but he almost always stops by in West Wardsboro for a brief or longer visit. He brings so many presents that I asked him to stop, but he insists and Terry and Kira really enjoy the elephants, tapestries, and incense. I lost touch with Bruce (until recently), and Judith is now severely incapacitated and no longer visits. They are still my now distant friends. Loyalty, caring, and commitment know no boundaries.

FOURTEEN
LOVERS, WIVES, AND DAUGHTERS

Love is a touchy subject. Everyone loves someone or something or some cat or canary. The word has very little real meaning especially in my family where people love or hate much too diversely. I reserve love for the few people whom I truly care about and hate for rapists, serial killers, and murderous tyrants like Caligula, Genghis Khan, Hitler, Stalin, and Mao Zedong. Lovers are a different story. From childhood on, and long before the onset of puberty, I liked females, a lot. I can recall with unequivocal clarity, and despite my segregation in an orthodox all-male Yeshiva, the young girlfriends my KV pals managed to acquire: Eleanor and Eva were very pretty. And Robin in high school, bohemian after my own heart. Lynn in college, said to be as promiscuous as some Ottoman oligarch. They may have been 10 or 15 or 20, but so was I, so my unfulfilled thoughts and desires were appropriately situated. In the eighth grade, I sat next to the much larger sixth and seventh—all in the same room. And it was here that Gladys, presumably 12 years old, could be found. She was extremely attractive, and as far out of my reach as Debbie Reynolds, who foolishly had chosen Eddie Fisher for a mate, for you see, Gladys's boyfriend had a car and was in the army! But all of this is of no consequence whatsoever for I was very shy and never once approached these (or other) girls with any premeditated intention. And so it went. As it happens, I have never outgrown my innate shyness; and I have never asked anyone for a formal date. I have always waited for good things to come to me. And they did.

One day, a friend mentioned that someone needed a place to stay. I obliged. Elena moved into the spare bedroom in my trailer, and went her own ways. She did not share costs; I was just being kind. She had no commitments but I was teaching and taking classes. We intersected very little. I once rhetorically asked her what she was planning to do on a Thursday or Friday night. She replied, *Why? Do you want a date?* and laughed hysterically. I never fully understood this, but shortly thereafter we decided to sleep together. That, naturally, did not work out well, and I still pay a hefty price for that blunder. Nevertheless, we both went to New York that summer, and shared an apartment. Elena must have missed Cleveland, because she returned there shortly after arriving in the City. Bizarrely, Sue, also jumped into bed with me and then wanted desperately to go to New York too, but I thought Elena was enough trouble so I declined her self-invitation. Perhaps that too was an error since she seemed to genuinely like sex.

So, there I was in Greenwich Village in 1967, the summer of love, alone, except for Susan (who is quite different from Sue). She and her sister Sandy lived in my parent's apartment building and I knew them in a casual sort of way. Through one of those amazing coincidences that occur with some frequency, Susan had gone to SUNY-Oswego, where she befriended Derora whom I met during my first year at Ohio. Somehow, we came together and once or twice enjoyed each other's pleasures. I gave her a key to my Jones Street apartment. One Sunday night, I returned from Vermont. I am a neat and meticulous person. I did not leave the drapes closed and the room littered with beer cans (perhaps because I never swallowed beer in my life). So, it was obvious that I had had a visitor in my absence. And here was a little note: Dear Bob, Thanks. I had a wonderful weekend with John O'Hara. Susan. Envy reared its ugly head. Why was Susan sleeping with O'Hara in my bed? What about me? And then I remembered. During the course of my life, I have known many people who read—a lot, including Bob and Terry, but I have only known two compulsive readers, people who cannot control their reading and do it at all times. A man in the post office who seemed to know every one of English's 500,000-plus words and Susan. She had spent three days reading O'Hara, perhaps *Appointment in Samara*. At least that would have been my choice.

I turned down a job with Schocken Books, where I might have met Samuel Agnon and Kafka, were he still alive. (I did meet Mr Schocken who expected me to read German and Hebrew and type, all for $75 a week.) Instead, I went for an interview at the post office (where the pay was much better). I saw a pretty girl with long hair. She was sitting alone far away. I went home. I came back a week later and there she was again. Hooray. We talked and it turned out that Phyllis, whom I rechristened Deirdre of the Sorrows, had come to the city from California or Nevada and lived on Staten Island. Hooray again, for I was going to pick up my little Alfa in the Village, because of the usual parking hassles, and take it to Staten Island. Would she like a ride? She said yes. So we took the subway from the General Post Office (GPO) down to Jones St, and went up to my diminutive apartment. (Years later, she told me that she was stunned that I didn't jump on her. How could I? I am very shy. What was she thinking, coming with a stranger to his apartment with such a suspicious attitude?) We talked and then we took the Ferry to my parents' house. My mom was home, and I introduced my new friend: *This is…* I had no idea what her name was. My mother made us lunch. So I had a girlfriend, sort of. Phyllis was not much enamored of sex, which (along with the acquisition of knowledge) has been the primary driving force throughout my adult life. Well, we did okay. We worked at the PO at night, went to Vermont once my parents got there, and sometimes she would stay at the Jones Street apartment. One day, my brother stopped by with my mail and Phyllis would not let him in and rightly so. He might have been an evil impostor. She was learning about tough city life. I hope that I got my letters.

I was not acting well because with Phyllis present in my next lovely apartment (on Bank St), I devoted my energies to a brand new friend, one who fascinated me because of her beauty and intelligence. After a while, we were doing things that you may read about in Henry Miller. Dana was a wonderful lover, but she sent me packing. I honestly do not know why. With two exceptions, Dana and I never did anything other than sleep together. She once asked me to walk with her to her therapist's office; I waited for an hour on the darkened street and then we walked home. And she wanted a wall cleared of plaster so that the underlying bricks would show to about three feet above the floor where wainscoting would complete the task. We went to a lumber yard, bought what we needed, and stripped the wall.

I put up the wood. She was very happy. Sometime afterwards, I met a man who lived in another building but with a contiguous wall. He mentioned that I should have warned him because our pounding vibrated his belongings off his shelves. I did not know him and had no way to get into his building, but still, I was (and am) sorry.

After I returned from Italy, in late 1967, and took the job with the Welfare Department, I met a young woman who sat nearby during the time that we spent in the office. Despite the fact that we lived together for a while, I never knew her real name! This is the only job I have ever had (or ever heard of) that required the presentation of a physical college diploma. Kathleen did not have one so she assumed someone else's identity (which must have wreaked havoc with her Social Security deductions). Although we worked deep in the heart of Brooklyn, we happened to live just a few blocks from each other in lower Manhattan. One day, I took a walk north past Tompkins Square (Needle) Park to Tenth Street and wandered into her hippie pad. I guess that she was disgusted with the inebriated or exhausted people lying on her floor, so she said, *Let's go out.* Her lovely, calm dog accompanied us. We went to my little apartment and sat around talking. Out of the blue, and with no precipitating indication, she announced, I'll sleep with you now. I am still amazed, almost 50 years later, because I had no inkling that this was an option. She really did dislike her own place and moved into a spare bedroom I had at another apartment. She lasted until I decided to move to yet another location, one that was indemnified by Marvin's mother, and thus I could not take a chance that something might go wrong. When I indicated that I was leaving and that she was not coming along, she began to howl like a lovesick coyote. I escaped. When I returned, she was gone and I never saw her again. She liked sex, but in surprisingly puritanical moderation.

In the spring of 1966, I was in the old Chubb library at Ohio listening to a recording of someone reading the complex poetry of Gerard Manley Hopkins. A young girl strolled up and asked if she could plug her earphones into the amplifier. I said yes. Terry was a sweet, ethereal 18 year-old, a freshman who liked poetry and theater. We enjoyed each other's company, wandering around the countryside and maybe seeing a movie but we never had a formal date. She came to visit me in New York during the 1967-1968 winter, and eventually we both ended up at Derora's country house in southeastern Ohio. We married in September. At first we lived in an apartment, but then were lucky enough to rent Jim Thompson's farm where we lived for my last two years of graduate study at OU. The farm was located about ten miles from Athens out in the rolling hills of southeastern Ohio. There were some woods but a lot of the land was devoted to pasturage on which we had cows and horses. Every morning, before I went to school, I would give grain and hay to the horses. They came to like me very much because I was very good to them. When they saw me they came running; when their various owners came out to ride, they ran away! We had osage orange, persimmon, and walnut trees—all very different from the eastern conifers I was used to; many species of birds; a pond; and a producing gas well that supported our heating, hot water, and cooking needs. It was as close to heaven as one is likely to get. That is why Jim and his wife decided to move there permanently.

Terry is a liberal; I am more radical (though not a leftist, but as far from the conservative/reactionary stance that my brother takes as a peacock is from a bacterium). She also perceives the world in a purposely irrational fashion. She sees angels dancing in trees and enumerates confusedly: One day recently, we went shopping together; she returned to the cart with her hands full. I said, I thought you were giving up cheese, and she replied, I only have one, while holding three. She had some truly bizarre explanation for what to a normal, numerate person would appear to be pure madness. All of this is especially hard for me, since I am a paradigmatically right-handed/left-brained thinker, logical and analytical, the very antipode of Terry, who has devoted her life to producing an exquisite body of poetry and paintings. She also returned to school late in life and earned a BA, MA, and PhD, which allowed her to teach in college. Interestingly, from the earliest days, we had an extremely open marriage. Once, I slept with a beautiful friend (whom I had long desired) and told Terry and she was not at all upset. Subsequently, she would occasionally sleep with someone other than me and I never said a word (unless she had been deceptive). But many years might go by without any extracurricular activity.

Figure 16: Terry in New Mexico, by Paula Chin
Figure 17: Terry and Bob, a Few Hours before Their Wedding, by Irving Herman

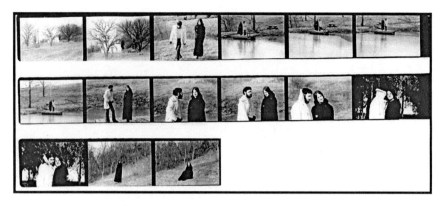

Figure 18: Fifteen Scenes from the Farm (originally shot for Seventeen, but never published), by Roger Wuchter

Then three things happened. We were living in Oklahoma and Terry had a lover who raised and sold dogs. One day, Terry was to fly to New York, and M asked her to take a dog along, which would save a fortune in shipping charges. I would take Terry to M's house but asked her to call first because we were to arrive at four AM and the door was open. But the house was filled with wild animals and I did not want them to tear Terry apart, but Terry is not afraid of anything so she refused to call, and we had a horrible fight. Well, somehow she survived and flew to the City. I was left behind in agonizing sadness. Appropriately, I worked on a paper dealing with Tadeusz Rozewicz's poetry. That night, I went to a friend's house for dinner. While Melissa cooked, two people showed up, and they decided to stay too. The man happened to live directly across the street from me, though I did not know him. I had never seen this gorgeous woman before. I sidled up to her and apologized for what we were about to eat (mushrooms), because my vegetarianism precluded chicken-fried steak, the normal Oklahoma fare. R muttered, I didn't expect to see you here, and, It's okay: I'm a vegan. I nearly fainted. The night wore on and R repeated, I didn't expect to see you here, many times. Well, she was a bit out of it but still... and apparently she had seen and admired me (she liked my beard) in the offices of *World Literature Today*, which I often visited, but I had not seen her. I do not understand this. It was now quite late, and I decided to leave. As I got to the door, she sat down at a piano and played the second fugue from the *Well-Tempered Clavier*. It is really impossible to indicate how I felt. If she had asked for my soul, I would have handed it over. Instead, I requested her phone number—in front of the other people. I wanted to call her at midnight (I should have), but I managed to hold off for a week or so, not wanting to appear overly anxious. We then met for lunch. After many, many hours, I thought that I really should return to work, and so should she. (I was a professor but she was a secretary.) I mentioned this and she surprised me: She had taken the afternoon off! She also mentioned that she had not thought that I would call, but she was very wrong there. We carried on for seven years. She was stunningly beautiful; like Angelina Jolie, she would have been a good companion for Brad Pitt at the Oscars; she was extremely intelligent; extremely talented as a musician and professional singer; and she liked sex, bizarre, outlandish, and frequent sex. I loved Terry, but somehow I was able to also love R at the same time—the only two adults I have ever fully and completely loved. I still care about both of them, although R has now disappeared despite my best efforts to remain connected. She brought joy to my heart and great pleasure. None of this caused any harm to Terry, who followed her own path. The only person who was hurt was R who wanted me to get divorced and marry her. I could not do that. I apologize. Shortly after the dog debacle, someone invited Terry to go away for a weekend with three other people. It did not take acute perceptivity to realize that this was a very precise setup, and I so informed Terry, but she, naturally, scoffed. (She always scoffs and I am always right.) She became the third person's lover and now, 30 years later, they are still in close contact.

One day, I was working at the library reference desk and two girls beset me with the following extraordinary remark: *Hi, Professor X suggested that we have an affair with you.* This is the kind of fantasy one may dream about as a teenager, but the reality was disconcerting, even at a time when no onus was attached to affairs with students. I thanked them for their consideration and they departed. But soon thereafter, one of them was back and again cajoling, inviting me to breakfast, working on me for no apparent reason, since there were many thousands of virile young men about all of whom would have been happy to accommodate Lynn. I fought back valiantly. Late one Saturday night, there was a knock on the door of my little house. Lynn sauntered in, all dressed up for seduction. I gave up. I never found out what this was all about, but she was true to her word: She had said that if I accommodated her, she would never bother me again. I only saw her twice after this: once in a supermarket and once, impossibly, in an otherwise empty library, when my Oklahoma friend R was visiting. R immediately guessed who she was. Yech! Most astonishing (and I am certain that some readers will disbelieve this) but ten or more years were to pass before it even dawned on me that I had slept with a 22 year old girl. I never gave any thought to my age, since I was youthful and conditioned. I was twice as old as she was.

And so matters stood. I do not recount every lover, every tryst that we managed during half a century, but there were many. Terry always had an excuse or regret, even when I warned her that what appeared to be occurring was merely a surface manifestation of hidden desire. She is too trusting and does not fully understand the power that sexual attraction and necessity hold over some people. Ten years after I had indicated that F was primarily interested in sleeping with her, he did. She moans and groans about how she made a mistake, but still pays little attention to my auguries. After R decided to give up and marry someone else, Terry went off to earn a Master's in New Mexico, and I was alone once again. I met a lovely young woman and eventually we became lovers. This too lasted for many years until she also decided to get married. Because of my youthful attitude and demeanor, I never thought, until later, about the age discrepancy, which was enormous. (I could easily have been her father.) We are still friends. Indeed, both Terry and I remain friends with many of our previous lovers, and they sometimes visit with us, or we with them. It is a strange situation but there you are.

I always have liked two diametrically different kinds of females: sweet, innocent, pleasant types and slutty, sexy vamps, who dress accordingly. If one likes short skirts and low-cut tops, one should not be embarrassed to associate with people who are willing to dress like that (to please me). Well, I managed as I could, cringing at times when the right person was in full regalia, and at a law conference where I spoke to a large audience composed of people (some of whom I knew from many years before) who would have been aghast had they seen or known or cared. I am now an old man, and no one wants a decrepit curmudgeon for a lover. Terry is much younger and vibrant. I will have to ask her how she is doing.

We had been married for 30 years, when Terry repeated for the hundredth time that she wanted to have a child. I was always very careful to avoid inducing pregnancy because I did not think that I would be a very good or competent parent. But now something was quite different. Terry was running out of time. So she was more forceful: If I do not have a child, I am leaving; and worse, I will be a bitter old woman. We went to an open adoption meeting in Minneapolis. Although Terry was still capable of having a child, I refused. Fifty year olds can die in childbirth and their offspring can be born with major health problems. So adoption seemed like a wiser choice. Terry hated the idea that the biological parents could visit at any time, so she investigated the more traditional route and after much work, travail, and financial outlay, she flew to Guatemala, on a spiritual quest, picked up 11 month old Kimberly (now Kira) and flew home. The entire trip took only four days. I met them at the Twin Cities airport and proceeded to drive the 80 Interstate miles to St Cloud in a rainstorm. They were both tired from the trip which led from Guatemala City to Fort Lauderdale to Houston to Minneapolis. It was very late at night. I drove in silence. About 50 miles from our destination, the rain turned to snow. I am an excellent snow driver and I always own four-wheel drive vehicles, so I continued rather than stopping at a motel. By Monticello, some 20 miles from St Cloud, and at midnight, the road became invisible. We were in a complete whiteout and alone; there were no other cars in either direction. It was impossible to tell where the road ended and the median began. In order to understand that I am not exaggerating, you should know that I have driven every type of vehicle (car, truck, forklift, large tractor, bulldozer) except a tank and semi-trailer. I have managed close to a million miles without an accident and I have driven through every type of horror. Even tornados have floated along beside me. I made it to my exit by placing my right-hand tires on the rumble strip and then slowly steering left, drifting, and then repeating this. I knew where I was because the big SUV rumbled. I was stressed out for a month. Kira never knew and now she is eleven. I am 60 years older than she is and we have a wonderful family: Terry, Bob, Kira, and Silver and Tiger. Kira tells me that the best thing to have happened to her, other than being adopted by us, is Silver, her diminutive but chubby cat, whom she adores. I know something about cats, and this one really is special: loving, docile, demanding, and playful. We are all very lucky. And no, life is not boring: We climb big mountains. Kira too.

FIFTEEN
CONCLUDING REMARKS

I began with some high-minded comments concerning an author's ethical commitment to facticity and his or her obligation to treat autobiographical writing as if it really were a true reflection of a life as it was lived. Memoir is not an opportunity or excuse to fictionalize one's existence. Naturally, one does not have to detail every indiscretion, and covert omissions are not the same as overt deceptions (not that I am making any excuses here; I have revealed much that others would have concealed). At the same time, I am neither Henry Miller nor Louis Ferdinand Céline, both of whom, among many others, have the ability and stomach to offer (in memoir or autobiographical fiction) revolting incidents and attitudes that a normal person would prefer to sublimate let alone reveal to a broad readership. My bathroom activities or personal prejudices (should I happen to have any) are none of anybody's business, and failing to describe them in detail is not an ethical breach.

But I'm not an innocent bystander and I cringe with embarrassment when, for example, I recall my early foolish attitude toward potential reproduction. I was conscious of and as a feminist cared enough about the other person to bring matters to the fore, something a randy male rabbit or (inebriated) human never does. But I commenced each session with a new lover by indicating that I was not responsible should she get pregnant. This now seems idiotic to me even though it was also my way of reminding her that pregnancy was a real possibility. (Who did I think was responsible?) Not a single person ever demurred and I wonder whether any of them ever bore a child (since I never used any form of protection and none of these people ever told me that she was doing so). This would be very sad because I have been denied a connection with my offspring. In one bizarre case, things had progressed to the point where we were in bed without any clothes on when the young woman calmly informed me that this would be the worst time of the month to have sex and she did not want to risk it. I respected her wishes and do not recall even feeling frustrated. Eventually, I stopped articulating my attitude toward potential parenthood, but not acting badly. Even after the first general indication, in around 1984, that sexual encounters could result in AIDS (and so scared people that a friend seriously stated that she would never have sex again), I did not use protection. This was unfair to both the other people and to me. The seriousness of all of this can be epitomized in the following devastating anecdote: In 1988, a gay friend told me that he was tired of going to funerals; this was his thirtieth. A few years later, this young man was forced to attend his own.

Some readers will undoubtedly view many of my actions in a negative light. That may be because they often accept and affirm social and cultural mores and political and religious dogma unthinkingly. Scrupulously objective consideration of many contemporary rituals, actions, beliefs, and even laws will indicate that they are misguided. I have chosen to ignore cultural mores, to go my own way but tried consistently to avoid harming others. I think, for the most part, that I have succeeded.

I have had an extraordinary life filled with joy and pleasure derived from helping others, as well as from sedentary pursuits such as reading, and extremely dangerous activities including high speed skiing and jumping, mountaineering in precarious locations in horrific weather, and solitary travel in stifling and dangerous deserts and killing blizzards. (I have never been more frightened than when walking up the half mile from our little southern Vermont town to the point where I parked the car in winter on a road I have traveled thousands of times since I was a child, but this day, the storm was so intense, I could hardly move. I did not think I could get to the woods and the protection of the car where I would recuperate before continuing on foot another half mile to our cabin.) I have flown through the air and landed on rocks, which would have broken another person's pelvis; I have gotten lost and spent many additional hours crawling my way out through deep snow, circumnavigating small cliffs, and using up the daylight. I have traversed extremely dangerous areas in cities all over the world, sometimes in the middle of the night (the Bowery in New York, the outer fringes of the Mission in San Francisco) and have survived.

Luck plays a big role here. I have managed to walk away from all of my injuries and encounters. I have stepped around bodies on the Bowery and know that people have been murdered there with some frequency. I have watched fellow gang members try to kill each other. I was lucky, at least in part, because I was very careful, and often deterred potential disaster by avoiding it: One night, Terry and I were walking on a New York east-west street (probably in the twenties) and I could see a major problem up ahead. Depending on the situation, were I alone, I might have continued (by crossing over to the opposite sidewalk), because I usually could outrun anyone. But since Terry was along, I suggested that we turn around, just as I quietly said to her, *Back out!* as we entered the Damascus gate, which leads into the Arab Quarter in Jerusalem's Old City. I knew we would be murdered, if we ventured on. (During our 1993 visit to Israel, at least one person was killed each day. Keep in mind that Israel is about the size of a typical Texas ranch and would disappear on a large privately held Australian station. Thus, this statistic is proportionally enormous.) Subsequently, I was told never to enter the Armenian Quarter, though I do not know why.

When the potential danger is imminent, one must simply move along, although there do exist instances when risking one's life to save someone is imperative; this, however, was not one of them: After a long, hard day at the law firm, I was almost at the Staten Island Ferry terminal when I encountered a rowdy crowd of stockbrokers, lawyers, secretaries, and other highly educated, white-collar professionals. They surrounded two derelicts who were attempting to kill each other probably with broken bottles. Instead of helping to avert disaster, these fools were egging the bums on, hoping to participate in calamity the way a person at a bullfight enjoys an animal's suffering. I would have lost had I intervened; this mob of lunatics was on the verge of hysteria and even the derelicts might have turned on me. I went home and never forgot the evil that is buried in these suburban parents' hearts.

The few negative experiences I have gone through, though

disconcerting and sometimes extremely emotionally or physically painful, have been so delimited, when contrasted with what depressives, kidnapping, rape, or assault victims, Holocaust survivors, or those afflicted with truly horrible maladies have suffered, that even when my pain is so intense and debilitating that I am given an extremely addictive narcotic (which years ago, some pharmacies did not stock, and which I refused to take, just as Marvin wisely refused to continue a morphine regime after a car accident), I comfort myself by thinking that in reality, I am truly blessed. Eventually, the pain subsides, and at 71, I can return to building or farming or serious mountaineering or breaking records: I am fairly certain that with a year of dedicated training, I could break the world record for the mile, in my age class, naturally. I will probably skip this though.

REFERENCES

This memoir is based almost exclusively on what I recall. But I have also superficially consulted some of my many letters. Whenever I was away from home for long periods of time, I wrote to my parents at least once a week. They saved these notes and gave them back to me. I also have the letters they sent me. Additionally, I have had ongoing literary and philosophical correspondence with many others including Bob and my father and brother. I have glanced at some of these as well. The Aiguille du Tour and Mt Mansfield accounts were originally published in the periodicals indicated below, under my name. Books (including my own) merely mentioned in passing are not included in this listing.

Douglas, William O. *Of Men and Mountains*. London: Victor Gollancz, 1951.

Hauptman, Robert. "Beaten on the Aiguille du Tour." *Altitude* 1.1 (April 1994): 10-12. (The version presented in *An Adventurous Life* has been emended.)

Hauptman, Robert. "Challenges on Mount Mansfield." *Long Trail News* 68.1 (Spring 2008): 31. (The version presented in *An Adventurous Life* has been slightly emended.)

Salzman, Mark. *Iron and Silk*. New York: Random House, 1986.

ABOUT THE AUTHOR

Robert Hauptman is a retired full professor. He worked as a reference librarian and an instructor for a quarter of a century, first at the University of Oklahoma and then at St Cloud State University in central Minnesota, where he taught undergraduate and graduate classes in information media and honors program courses in the humanities and social sciences. He holds a BA in German, MA in English, MLS in library science, PhD in comparative literature, and PhD (ABD) in library science. He has some 600 publications in four disciplines: literary criticism, library science, ethics, and general interest. He is the co-author of *The Mountain Encyclopedia* (2005) and *Grasping for Heaven: Interviews with North American Mountaineers* (2011); his latest scholarly studies are *Documentation* (2008) and *Authorial Ethics* (2011); and he is the founding and current editor of the *Journal of Information Ethics*. He continues to write and climb. During the summer of 2005, he climbed 19 times including Washington's Baker and Adams; in the 2006 season he climbed Rainier (though he did not reach the summit of these western peaks on these trips). In 2012, he did Lassen, the South Sister, and Adams. He climbs Vermont mountains, e.g., Stratton, Mt Snow, and Mansfield, very frequently. And he has stood on the high points of 45 of the 50 states. He is 71 years old and has an 11 year old daughter; Kira likes to climb too.

OTHER ANAPHORA LITERARY PRESS TITLES

Interviews with Best-Selling Young Adult Writers
Editor: Anna Faktorovich

East of Los Angeles
By: John Brantingham

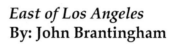

Death Is Not the Worst Thing
By: T. Anders Carson

Folk Concert
By: Janet Ruth Heller

100 Years of the Federal Reserve
By: Marie Bussing-Burks

River Bends in Time
By: Glen A. Mazis

Interviews with BFF Winners
Editor: Anna Faktorovich

Compartments
By: Carol Smallwood

CPSIA information can be obtained
at www.ICGtesting.com
Printed in the USA
EDOW021306150313
926ED